BURLEIGH DODDS SCIENCE: INSTANT INSIGHTS

NUMBER 118

Land use change and management

I0130563

burleigh dodds
SCIENCE PUBLISHING

Published by Burleigh Dodds Science Publishing Limited
82 High Street, Sawston, Cambridge CB22 3HJ, UK
www.bdspublishing.com

Burleigh Dodds Science Publishing, 1518 Walnut Street, Suite 900, Philadelphia, PA 19102-3406, USA

First published 2025 by Burleigh Dodds Science Publishing Limited
© Burleigh Dodds Science Publishing, 2025. All rights reserved.

British Library Cataloguing in Publication Data
A catalogue record for this book is available from the British Library

ISBN 978-1-83545-258-5 (Print)
ISBN 978-1-83545-259-2 (ePub)

DOI: 10.19103/9781835452592

Typeset by Deanta Global Publishing Services, Dublin, Ireland

Contents

Series list

Title	Series number
Sweetpotato	01
Fusarium in cereals	02
Vertical farming in horticulture	03
Nutraceuticals in fruit and vegetables	04
Climate change, insect pests and invasive species	05
Metabolic disorders in dairy cattle	06
Mastitis in dairy cattle	07
Heat stress in dairy cattle	08
African swine fever	09
Pesticide residues in agriculture	10
Fruit losses and waste	11
Improving crop nutrient use efficiency	12
Antibiotics in poultry production	13
Bone health in poultry	14
Feather-pecking in poultry	15
Environmental impact of livestock production	16
Sensor technologies in livestock monitoring	17
Improving piglet welfare	18
Crop biofortification	19
Crop rotations	20
Cover crops	21
Plant growth-promoting rhizobacteria	22
Arbuscular mycorrhizal fungi	23
Nematode pests in agriculture	24
Drought-resistant crops	25
Advances in detecting and forecasting crop pests and diseases	26
Mycotoxin detection and control	27
Mite pests in agriculture	28
Supporting cereal production in sub-Saharan Africa	29
Lameness in dairy cattle	30
Infertility and other reproductive disorders in dairy cattle	31
Alternatives to antibiotics in pig production	32
Integrated crop-livestock systems	33
Genetic modification of crops	34

Acknowledgements

Chapters in this Instant Insight are taken from the following sources:

Chapter 1 The role of agricultural expansion, land cover and land-use change in contributing to climate change
Chapter taken from: Deryng, D. (ed.), Climate change and agriculture, Burleigh Dodds Science Publishing, Cambridge, UK, 2020, (ISBN 978 1 78676 320 4; www.bdspublishing.com)

Chapter 2 Understanding how land-use management affects soil microbiomes
Chapter taken from: Dunfield, K. (ed.), Understanding and utilising soil microbiomes for a more sustainable agriculture, Burleigh Dodds Science Publishing, Cambridge, UK, 2025, (ISBN 978 1 80146 474 1; www.bdspublishing.com)

Chapter 3 Implementing sustainable land use change programmes
Chapter taken from: Reid, N. and Smith, R. (eds.), Managing biodiversity in agricultural landscapes: Conservation restoration and rewilding, Burleigh Dodds Science Publishing, Cambridge, UK, 2024, (ISBN 978 1 80146 454 3; www.bdspublishing.com)

Chapter 4 Agroforestry practices: riparian forest buffers and filter strips
Chapter taken from: Mosquera-Losada, M. R. and Prabhu, R. (eds.), Agroforestry for sustainable agriculture, Burleigh Dodds Science Publishing, Cambridge, UK, 2019, (ISBN 978 1 78676 220 7; www.bdspublishing.com)

Chapter 1

The role of agricultural expansion, land cover and land-use change in contributing to climate change

Catherine E. Scott, University of Leeds, UK

1 Introduction

Land-use change has accompanied the arrival and movement of human populations into and between regions of the Earth for thousands of years. The dominant effect of human arrival is the removal of forests to provide land that can be used for agriculture. Reliable land-use surveys exist only from the mid-twentieth century onwards, so the Earth's vegetation distribution must be reconstructed prior to that. On geological time scales, the knowledge of climatic conditions and indicators in the fossil record enable a reconstruction of natural vegetation across the globe.

Reconstructing land cover during the period of more substantial human influence (i.e. the past several thousand years) presents considerable challenges and is often based on estimates of the human population as well as the assumption that the amount of land 'used' per person for agricultural purposes has remained broadly similar over time (e.g. Ramankutty and Foley 1999; Klein Goldewijk 2001; Pongratz et al. 2008; Klein Goldewijk et al. 2011).

An alternative approach (Boserup 1965; Ruddiman 2003; Kaplan et al. 2009) considers the possibility that the amount of land required per capita may have declined substantially over time due to the intensification of agricultural

http://dx.doi.org/10.19103/AS.2020.0064.10

practises. Combining this approach with the estimates of population change results in much larger areas of land being under agricultural use over the past two millennia (Fig. 1; Kaplan et al. 2011).

Up until the Industrial Revolution in the eighteenth century, there was widespread deforestation across the temperate regions of Asia, Europe and North America on the land that was considered most suitable for farming (Williams 2003). At the start of the twentieth century, deforestation rates in tropical regions began to accelerate (FAO 2012), particularly in South and Central America, Southeast Asia and Central Africa.

Forest clearance in the tropics still occurs predominantly to acquire the land that is suitable for agriculture with more than 80% of new agricultural land acquired across the tropics between 1980 and 2000 coming from the clearance of intact or disturbed forests (Gibbs et al. 2010); however, the specific commodities driving agricultural clearance vary from region to region within the tropics.

In South and Central America, beef cattle ranching has been the dominant driver of 'modern' forest clearance (e.g. Grainger 1993; Fearnside 2005; Gibbs et al. 2010). During the 1960s, 1970s and 1980s, road building and financial incentives by the Brazilian government encouraged deforestation of the Amazon to create pasture and cattle ranches (Carvalho et al. 2002; Fearnside 2005). This process often involves clearance of the forest followed by burning

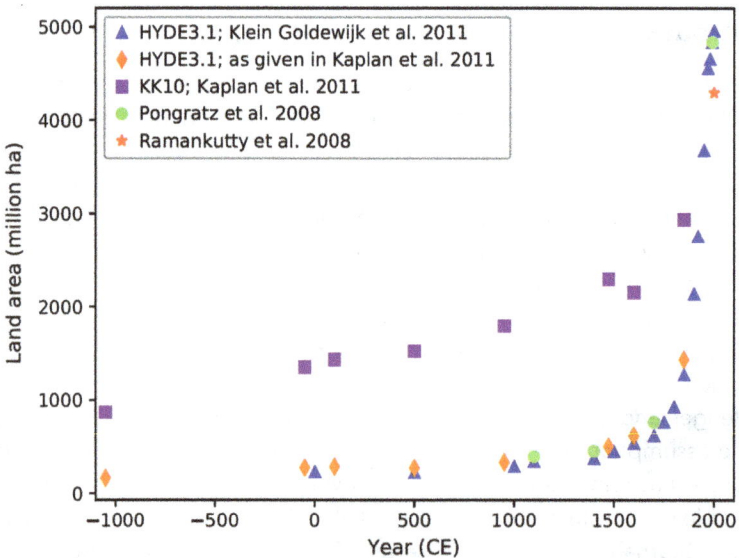

Figure 1 Land area occupied by agricultural activities (i.e. cropland and pastureland combined) in historical reconstructions up to the present day (Pongratz et al. 2008; Ramankutty et al. 2008; Kaplan et al. 2011; Klein Goldewijk et al. 2011).

to remove any residual trees. Until the 1990s, Brazilian beef was usually sold domestically, but in the early 2000s, international demand for Brazilian beef partly drove a spike in deforestation rates between 2002 and 2004 (Nepstad et al. 2006).

During the 1970s, an oil embargo prompted a rapid expansion of biofuel crop growth in South America, specifically the growth of sugar cane to produce ethanol. In 1977, the Brazilian government mandated that all gasoline must be blended with ethanol. This mandate is still in place and the current minimum blend level is set at 27% ethanol. As well as directly driving forest clearance, the growth of bioenergy crops can indirectly lead to deforestation if it displaces food production, which then moves onto forested land.

During the 1990s and 2000s the growth of soybeans also began to contribute substantially to the clearance of Amazon forests. Rather than being directly consumed by humans, soybean crop is mainly used to feed cattle, pigs and chicken. In the late 1990s, new cultivars enabled farmers to grow soybeans in regions that had not previously been climatically suitable, leading to a rapid expansion of soy farms into the Amazon forest (Fearnside 2001; Nepstad et al. 2006). In response to growing environmental concerns, a moratorium was announced by the exporters and processors of soybeans stating that they would not buy crops grown on farmland within the Brazilian Amazon that had been deforested since June 2006. Since its implementation, the soy moratorium appears to have been successful, with most new soy expansion occurring on previously cleared land (Rudorff et al. 2011; Gibbs et al. 2015) and has contributed to the overall decline in Brazilian deforestation rates (Hansen et al. 2013).

In Southeast Asia, extensive forest loss has been driven by food and fuel crop growth, as well as rubber and timber production (Gibbs et al. 2008; Miettinen et al. 2011). Malaysia and Indonesia now produce more than 80% of the world's palm oil (USDA-FAS 2019) through both industrial scale and smallholder plantations (Schoneveld et al. 2019). Oil palms grow only in humid, tropical conditions but are extremely efficient producers of oil compared to other crops (e.g. soybean, sunflower, rapeseed). The major environmental issue associated with oil palm growth is the conversion of old-growth or peat forests which contain dense carbon stocks, 6% of tropical peatlands in Southeast Asia had been converted to oil palm plantations by the early 2000s (Koh et al. 2011). In 2011, the Indonesian government imposed a moratorium on new oil palm and timber plantations on peatlands or primary forests, but its effectiveness remains unclear (Busch et al. 2015).

In Africa, forests are cleared to provide wood fuel and to make way for smallholder agriculture, but information about the scale and extent of deforestation is less robust than for South America or Southeast Asia, both due to a lower reporting capacity and the challenges associated with detecting small-scale forest clearance by satellites (Malhi et al. 2013). The lack of industrial-scale land clearance for agriculture means that deforestation rates in Africa have remained lower than those in South America or Southeast Asia throughout the 1990s and 2000s (Achard et al. 2002; Mayaux et al. 2013). Other pressures on tropical forests include mineral mining (e.g. to obtain gold, copper, tin), coal mining and oil drilling which have been particularly prevalent in parts of South America and Africa (Grainger 1993).

Monitoring changes to land cover and rates of forest loss relies on either country-level statistics of forest area (e.g. FAO 2018) or, in recent years, remote sensing from both airborne (e.g. Asner 2009; Saatchi et al. 2011a) and satellite instruments (e.g. Defries et al. 2000; Hansen et al. 2003, 2013; Saatchi et al. 2011b). At the global scale, natural forests were being lost at a rate of 10.6 million ha per year during the 1990s. Between 2010 and 2015 this rate slowed to approximately 6.5 million ha lost per year (FAO 2015). For the 2010-2015 period, the overall rate of forest area change was estimated as a net loss of 3.3 million ha per year (FAO 2015); this is lower than the rate of direct forest loss due to extensive afforestation, particularly in China (Fang et al. 1998, 2001; Wang et al. 2007), and the natural expansion of forests onto previously managed lands.

2 Impacts of land-use change on climate

When land cover or land use is changed, the fluxes of carbon, energy and water between the land-surface and the atmosphere can be altered substantially (Fig. 2; Bonan 2008). Changes to these fluxes can be broken down into radiative (i.e. energy) and non-radiative (i.e. carbon and water) effects. The following sections outline our understanding of the way that these fluxes change, and estimates of the extent to which this has occurred due to agricultural expansion.

2.1 Carbon emission

Plants take in carbon, in the form of carbon dioxide (CO_2), from the atmosphere. Approximately half of this carbon is returned to the atmosphere during respiration, while the other half is fixed as plant biomass during photosynthesis. The metabolic activity of ecosystems influences the amount of carbon in the atmosphere, with a marked seasonal cycle evident in measurements of atmospheric CO_2 concentrations.

The total amount of carbon stored in terrestrial ecosystems is uncertain: the tropical forests of Latin America, sub-Saharan Africa and Asia are estimated to contain between 247 and 553 petagrams of carbon (PgC), temperate forests between 159 and 292 PgC and boreal forests between 395 and 559 PgC (Dixon et al. 1994; Prentice et al. 2001; Saatchi et al. 2011b). As a result of the carbon stored in the trees, a considerable emission of carbon can be associated

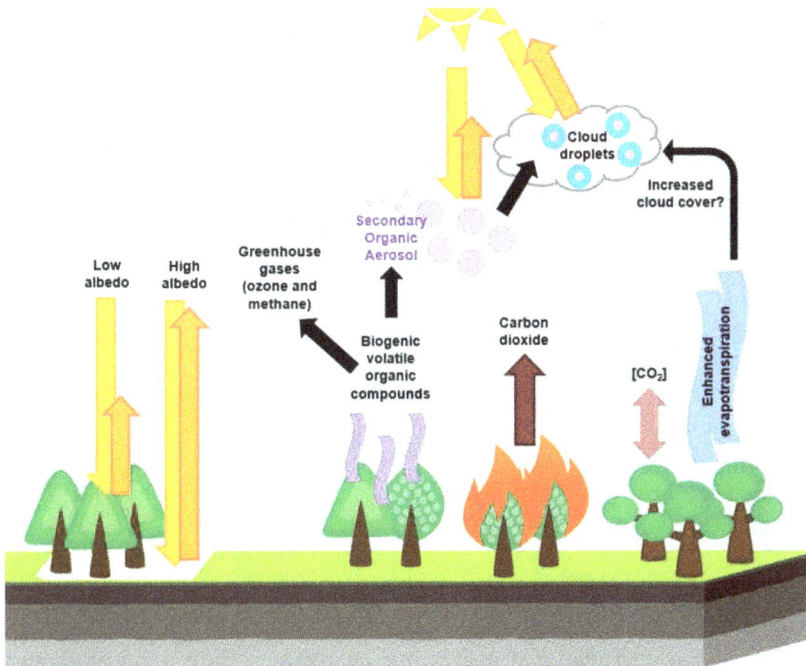

Figure 2 Schematic of the processes described in Section 2 of this chapter.

with the process of land-use and land cover change (LULCC), predominantly through the decay and burning of vegetation when forests are converted to agricultural land. Conversely, the process of afforestation will remove carbon dioxide from the atmosphere, generating a carbon sink.

Estimating the net flux of carbon to the atmosphere as a result of LULCC is challenging. Three methods are often used: (1) a 'bookkeeping' method that combines country-level forest area data with regionally averaged biomass values (e.g. Houghton 2003; Baccini et al. 2012); (2) dynamic global vegetation models (DGVMs) that simulate the fluxes of carbon between the land surface and the atmosphere; and (3) regional analyses based on satellite data (e.g. van der Werf et al. 2010).

While fluxes of carbon within the carbon cycle are conventionally referred to as an amount of carbon 'C', anthropogenic emissions are more commonly referred to in terms of a quantity of carbon dioxide 'CO_2', the value of which will be a factor of 3.67 (44 ÷ 12) higher for the same amount of carbon. For the most recent decade available (2008–2017), annual emissions from LULCC are estimated to be 1.5 ± 0.7 PgC (equivalent to 5.5 $PgCO_2$; Le Quéré et al. 2018), approximately 14% of the total annual carbon emission from anthropogenic activities. Despite this annual source of carbon emission being generated due to LULCC, globally the land remains a carbon sink because vegetation removes more carbon dioxide from the atmosphere than that is put back (Jia et al. 2019).

The challenges posed by quantifying LULCC emissions in the present day are further exacerbated prior to the mid-twentieth century where the changes in land-cover must be inferred (see Section 1). It is estimated that since the year 1750 CE, the process of land-use change has resulted in a total emission of almost 200 PgC into the atmosphere, compared to an estimated 27 PgC emitted due to LULCC between 800 CE and 1750 CE (Pongratz et al. 2009; Ciais et al. 2013). However, Kaplan et al. (2011) suggest that over 300 PgC could have been emitted as a result of agricultural land clearance by 1850 CE due to the larger area occupied in their estimates (Fig. 1).

2.2 Surface energy fluxes

2.2.1 Reflection of solar radiation

As well as influencing the atmospheric concentration of CO_2, the presence of large-scale vegetation also affects the energy balance at the Earth's surface. Forests are generally darker in colour than other land surface types, particularly cultivated vegetation such as cropland. This dark surface means that forested land absorbs most of the shortwave radiation that it receives from the Sun. The ratio of reflected to incident shortwave solar radiation is known as the albedo. A very bright surface, for example, fresh snow, would have an albedo of around 0.9, whereas dark surfaces like the ocean have a much lower albedo of around 0.1. Forests typically have a very low albedo of 0.08-0.19 (Betts and Ball 1997; Monteith and Unsworth 2008), whereas grass or cropland have a slightly higher albedo of 0.15-0.26 (Monteith and Unsworth 2008).

In the boreal region (above 60°N), snow covers the land surface for several months of the year. If this snow cover is lying on short vegetation, it will completely cover it and the surface albedo will be very high, for example, 0.75 (Betts and Ball 1997). However, if coniferous boreal trees are present, they will protrude from the snow, lowering the albedo to 0.1-0.15 (Leonard and Eschner 1968; Robinson and Kukla 1985; Thomas and Rowntree 1992; Betts and Ball 1997).

For an evergreen forest in a snow-free region, the albedo will be relatively constant year round. However, for the deciduous forests occupying temperate and boreal regions, the albedo of the forest varies according to the time of year and whether or not the trees are in leaf. In the winter months, the albedo of a deciduous forest will be mainly controlled by the underlying surface, and the process of leaf-out tends to increase the albedo of a deciduous forest by 20-50% (Hollinger et al. 2010; Richardson et al. 2013).

The overall climate impact of land-cover change is explored in Section 3, but the conversion of forest to cropland or other agricultural land will generally result in an increase in the surface albedo. The LULCC that has occurred since 1750 is therefore considered to have increased the overall albedo of the Earth's land, resulting in an overall cooling effect on the climate. This is quantified as a radiative forcing of approximately -0.18 W m^{-2} (Myhre et al. 2013).

2.2.2 Evapotranspiration and hydrological impacts

The presence of vegetation also mediates the transfer of water from the land surface to the atmosphere via evapotranspiration. Evapotranspiration is the sum of physical evaporation from soils and surfaces in the canopy, and biological transpiration. During transpiration, water that has been taken up from the soil via plant roots is lost through the stomata on leaves.

Higher rates of evapotranspiration have been measured above forests than other land-cover types (Spracklen et al. 2018). During the dry season, trees are able to sustain high evapotranspiration rates because their long roots, when compared to other vegetation, facilitate access to deep soil water (Nepstad et al. 1994; Canadell et al. 1996). Evapotranspiration plays such a strong role in hydrological cycling that parcels of air travelling over forests have been shown to produce at least twice as much rainfall as the air that has passed over little vegetation (Spracklen et al. 2012).

By modulating water fluxes, forests may also alter the distribution of low-level clouds. Observational studies have reached conflicting conclusions on the impact of deforestation on cloud cover. Wang et al. (2009) found that shallow clouds formed preferentially over patches of land in the Amazon that had been deforested, while Teuling et al. (2017) saw a strong increase in cloud cover over forested regions of western Europe.

As well as being an important part of the hydrological cycle, the process of evapotranspiration plays a vital role in the Earth's surface energy balance. The energy associated with a change from liquid water into water vapour during evapotranspiration is referred to as latent heat. The other important energy flux from the surface to the atmosphere is sensible heat, which refers directly to the change in atmospheric temperature. The ratio of sensible to latent heat is known as the Bowen ratio (Bowen 1926); by controlling evapotranspiration, vegetation can affect the Bowen ratio and influence local temperatures. Conversion of forests to cropland or grassland is therefore likely to reduce evapotranspiration rates, altering the Bowen ratio and potentially increasing local temperatures.

2.3 Emission of reactive gases from vegetation

In addition to storing carbon, controlling the reflectivity of the land surface and mediating the transfer of moisture to the atmosphere, vegetation present on the land surface can influence the atmospheric concentrations of a number of climatically important non-CO_2 greenhouse gases and particles.

The vegetation emits biogenic volatile organic compounds (BVOCs) into the air and the BVOCs include isoprene (with chemical formula C_5H_8), monoterpenes ($C_{10}H_{16}$) and sesquiterpenes ($C_{15}H_{24}$), with their emission rates dependent upon plant species, temperature, sunlight levels and atmospheric CO_2 concentration. Approximately 500 Tg of isoprene is emitted annually by vegetation, with around 100 Tg of monoterpenes and 30 Tg of sesquiterpenes (Went 1960; Rasmussen and Went 1965; Sanadze and Kursanov 1966; Guenther et al. 1991, 2012).

Producing BVOCs requires a large investment of energy from plants. This investment suggests that there is some form of advantage to be gained by their emission. Potential benefits to the plant include: enhancing resilience abiotic stress (e.g. temperature, light and oxidative damage; Loreto and Velikova 2001; Vickers et al. 2009), preventing the establishment of competing plants (Muller 1966), altering the climate (e.g. temperature, precipitation and cloud cover; for example, Spracklen et al. 2008; Paasonen et al. 2013; Scott et al. 2018a), allowing belowground signalling (e.g. Rasmann et al. 2005), or reducing insect and herbivore attack (e.g. Oh et al. 1967; Kessler and Baldwin 2001; Amin et al. 2013).

Once emitted into the atmosphere, BVOCs undergo a series of chemical reactions to give a wide range of products. One consequence of these atmospheric reactions is the formation of biogenic particles. By scattering incoming solar radiation, and acting as seeds for cloud droplet formation (thereby increasing the brightness of clouds), biogenic particles are likely to have a cooling effect on the global climate (Scott et al. 2014; Zhu et al. 2017, 2019). Recent research suggests that the products of monoterpene oxidation may be involved in the very first stages of new particle formation in the atmosphere, a process that was previously thought to rely on the presence of human-made pollution (Gordon et al. 2016; Kirkby et al. 2016).

Due to the complex atmospheric chemistry in which BVOCs participate, their presence can alter the concentration of some non-CO_2 greenhouse gases (ozone (O_3) and methane (CH_4)) which have a warming effect on the climate (Unger 2014). When evaluated over a 100-year time period, CH_4 is estimated to be between 28 and 34 times more effective at warming the climate than CO_2 (Myhre et al. 2013). The ability of O_3 to warm the climate is dependent on its location in the atmosphere, so it is not possible to quantify in quite the same way as CH_4. However, the rise in O_3 concentrations in the lower atmosphere due to anthropogenic activity is thought to be the third largest contributor to climate change of all greenhouse gases (behind CO_2 and CH_4; Myhre et al. 2013).

The results of the chemical reactions that BVOCs participate in depend upon the concentration of other gases in the atmosphere. While BVOCs can react directly with O_3, decreasing its concentration, under certain conditions (in the presence of sufficient nitrogen oxides), the emission of BVOCs contributes to the production of O_3, leading to an overall increase in O_3 concentration (Monks et al. 2015).

Unger (2014) found that LULCC since the year 1850 had resulted in a net cooling effect on climate through decreases in BVOC emission and therefore O_3 and CH_4 concentrations. In contrast, Scott et al. (2018b), found that the warming effect associated with a reduction in biogenic particles outweighed the cooling effect due to a reduction in O_3 and CH_4, leading to a 10% enhancement of the overall warming due to deforestation.

3 Estimating the impacts of land-use change on climate

The climate impact of land-use change is difficult to isolate from observations, so computer models can be used to explore the effects of converting one land cover type to another in idealised experiments (e.g. Lean and Warrilow 1989; Bonan et al. 1992; Betts 2000; Claussen et al. 2001; Bounoua et al. 2002; Snyder et al. 2004; Feddema et al. 2005; Gibbard et al. 2005; Bala et al. 2007; Davin and de Noblet-Ducoudré 2010; Pongratz et al. 2010; Arora and Montenegro 2011; Swann et al. 2012; Hallgren et al. 2013).

Modelling studies find that the overall impact of deforestation on climate is latitude dependent. Bala et al. (2007) found that the net climate impact of simulated global forest removal was a temperature reduction of −0.3°C. Removing tropical forest led to a global mean warming (+0.7°C) due to a reduction in evapotranspiration and high carbon storage in the tropics, whereas the removal of boreal forests resulted in a global mean cooling (−0.8°C) due to the dominance of the surface albedo effect. Davin and de Noblet-Ducoudré (2010) also found that, when looking only at the biogeophysical impacts, simulated global forest removal generated a cooling (−1°C) with the albedo effect dominant at high latitudes, and the effects of reduced surface roughness and evapotranspiration dominant in the tropics.

In temperate regions, the balance between competing biogeophysical effects is less clear than for either boreal or tropical forests (Claussen et al. 2001). Bala et al. (2007) simulated a global annual mean cooling of −0.04°C for total temperate deforestation. For the northern hemisphere (NH) alone, Snyder et al. (2004) obtained a larger annual mean cooling of −1.1°C and found that the effect was seasonally dependent, with temperate deforestation causing a cooling during the local winter and a warming during the summer. Swann et al. (2012) found a global mean temperature change of between −0.4°C and +0.1°C due to northern mid-latitude afforestation, but also simulated a northward shift of tropical precipitation belts and drying of the southern Amazon.

All the previously mentioned studies used models to examine idealised deforestation or afforestation scenarios. Estimating the climatic impact of historical land-use change combines the challenges associated with understanding the climate impact of specific land-cover transitions, with the challenge of reconstructing historical land-use change. Pongratz et al. (2010) estimated that LULCC since 1850 has resulted in a biogeochemical warming of 0.16–0.18°C and biogeophysical cooling of −0.03°C, giving a combined overall warming.

4 Role of the land sector in climate change mitigation

The 2015 Paris Agreement on Climate commits signatories to 'hold the increase in the global average temperature to well below 2°C above pre-industrial levels' and to 'pursue efforts to limit the temperature increase to 1.5°C'. Globally, anthropogenic CO_2 emissions are still rising, and emissions from fossil fuel combustion have increased from 3.1 ± 0.2 PgC (11.4 $PgCO_2$) per year during the 1960s to an average of 9.4 ± 0.5 PgC (35 $PgCO_2$) per year between 2008 and 2017 (Le Quéré et al. 2018). Future emission projections that are able to limit the global temperature increase to 1.5°C tend to require that global CO_2 emissions peak around 2020 and reach net zero by mid-century (Rogelj et al. 2018).

Achieving net zero emissions means that any remaining CO_2 emissions are balanced by processes that remove CO_2 from the atmosphere. This can be achieved in a number of ways, but the most frequently cited strategies are an increase in afforestation or reforestation and the large-scale deployment of bioenergy with carbon capture and storage (BECCS). BECCS involves the deliberate growth of crops that can be burned to generate energy, and the subsequent burial of the CO_2 emitted during their combustion; estimates suggest that BECCS could remove around 3 PgC (11 $PgCO_2$) per year by 2100, requiring up to 0.7 billion ha of land (Smith et al. 2016).

By 2050, the world's population is projected to reach 9 billion and the Food and Agriculture Organisation predicts that a 70% increase in food production will be required (Food and Agriculture Organisation of the United Nations (FAO) 2009). This is unlikely to be achieved by improving agricultural yields or intensifying livestock production alone, suggesting that the expansion of the current agricultural land area would occur (Bajzelj et al. 2014). Taking the requirements for food and bioenergy production together gives a complex set of possible future land-use change scenarios; a combination of increasing population, potentially increasing meat consumption and a requirement for BECCS will place enormous demands on global land that may not be sustainable (Benton et al. 2018). Over 75% of current agricultural land is used to raise livestock; future dietary changes, such as a reduction in meat consumption, therefore have the potential to reduce the amount of land required (Stehfest et al. 2009), as well as the greenhouse gas emissions associated with agriculture (Springmann et al. 2016).

In conjunction with the Paris Agreement on Climate, countries around the world prepared Nationally Determined Contributions (NDCs) to state what they would do to mitigate, and adapt to, future climate change. Initial assessments of the NDCs suggest that countries are currently expecting approximately one-quarter of their mitigation targets by 2025–2030 to be met by the land-use sector, through reduced deforestation and increased afforestation (Forsell et al. 2016; Grassi et al. 2017).

4.1 Reducing deforestation

The UN-REDD (Reducing Emissions from Deforestation and forest Degradation) programme, and the extension REDD+, aims to reduce forest loss in developing countries by introducing financial mechanisms to benefit countries that preserve the carbon stocks in their forests (e.g. Angelsen and Wetrtz-Kanounnikoff 2008). Reducing deforestation rates by 50% by 2050 (relative to rates observed in the 1990s), and maintaining them at that level until 2100 would avoid the direct release of approximately 50 PgC (Gullison et al. 2007), equivalent to five years of fossil fuel carbon emissions. Using a dynamic global vegetation model, Gumpenberger et al. (2010) found that tropical carbon stocks in 2100 decreased by 35 and 134 PgC, relative to 2012, under a continued deforestation scenario, whereas under a forest protection scenario tropical carbon stocks could be increased by between 7 and 121 PgC.

4.2 Increasing reforestation, restoration and afforestation

While preserved or increased forest cover would enhance CO_2 sequestration and storage, forests also exert the biogeochemical and biogeophysical impacts discussed in sections 2 and 3; as such the overall climatic impact of modifications to forest area will be complex and location specific.

Pongratz et al. (2011) found that the majority of historical anthropogenic LULCC in temperate and boreal regions has occurred on the most productive land, thereby generating higher than average (i.e. for a particular latitude) CO_2 emissions. Subsequently, reforestation of these areas could potentially induce a cooling effect from CO_2 sequestration that would outweigh any warming effect due to an albedo increase. Arora and Montenegro (2011) found that gradually replacing cropland in an Earth system model with forests reduced the simulated global mean temperature at the end of the twenty-first century by 0.45°C because the impact of increased carbon sequestration outweighed the warming from biogeophysical effects.

Using photo-interpretation of satellite imagery, Bastin et al. (2019) identified 0.9 billion ha of land globally that could support forests. Their analysis suggested that these additional forests could potentially store 205 PgC, but did not consider the biogeophysical impacts of the additional forests and may have overestimated the capacity for above-ground and soil carbon increases (Friedlingstein et al. 2019; Lewis et al. 2019; Veldman et al. 2019).

4.3 Growth of crops for bioenergy

Since vegetation takes in CO_2 during photosynthesis (see Section 2.1), it is theoretically possible to generate carbon-neutral energy by harvesting and burning crops and other plants. This is often done using fast-growing perennial grasses such as Miscanthus (also known as silvergrass) or coppicing short-rotation trees such as poplar, willow and eucalyptus. While the use of bioenergy crops can potentially displace fossil fuels and therefore reduce carbon emissions, bioenergy crops grown on former high carbon forest or peatland may struggle to repay the carbon debt associated with their initial establishment (Harper et al. 2018). If combined with carbon capture and storage (CCS) the global mitigation potential of BECCS is estimated to be up to around 3 PgC per year (Smith et al. 2016; Jia et al. 2019).

While motivated by the need to reduce carbon emissions, the large-scale growth of bioenergy crops could also have substantial impacts on surface energy fluxes and therefore local climate. Modelling studies indicate that the expansion of perennial bioenergy crops onto land previously used to grow annual crops would lead to an increase in both surface albedo and evapotranspiration, and therefore a localised cooling effect (Georgescu et al. 2011).

Many bioenergy crops emit higher levels of isoprene than the food crops they may have replaced. Ashworth et al. (2012) found that, while the impacts on global climate were negligible, replacing food crops with oil palm and short-rotation coppice resulted in localised increases in both O_3 and secondary organic aerosol concentrations.

5 Future land-use trajectories

As discussed in Section 3, computer models are often used to explore the impact of changes in land cover on the climate. The same approach is taken to assess future potential climate change as a result of different levels of greenhouse gas, and other anthropogenic, emissions.

A set of Shared Socioeconomic Pathways (SSPs; Riahi et al. 2017) has been developed that describe five different future narratives for society. The SSPs reflect a range of levels of possible global challenges around mitigation and adaptation to climate change (Table 1), but do not include

Table 1 Summary of the Shared Socioeconomic Pathways (Riahi et al. 2017) and their implications for future land-use change (Popp et al. 2017)

Shared Socioeconomic Pathway	Overview of pathway	Description of agriculture and land-use sector
SSP1	Sustainability: Taking the Green Road (*low challenges to mitigation and adaptation*)	Strong land-use regulation. Improvements in agricultural productivity, low meat diets and low growth in food consumption. Full participation of land-use sector in mitigation.
SSP2	Middle of the Road (*medium challenges to mitigation and adaptation*)	Medium land-use regulation. Material-intensive consumption and medium meat consumption. Slow decline in deforestation rate. Partial participation of land-use sector in mitigation.
SSP3	Regional Rivalry: A Rocky Road (*high challenges to mitigation and adaptation*)	Limited regulation. Resource intensive consumption. Continued deforestation. Limited participation of land-use sector in mitigation.
SSP4	Inequality: A Road Divided (*low challenges mitigation but high challenges to adaptation*)	Uneven regulation. High deforestation rates in low-income countries. Unequal consumption. Partial participation of land-use sector in mitigation.
SSP5	Fossil-fuelled Development: Taking the Highway (*high challenges to mitigation but low challenges to adaptation*)	Medium regulation. Material-intensive consumption and meat-rich diets. Slow decline in deforestation rate. Full participation of land-use sector in mitigation.

specific climate policies. Integrated assessment models are then used to realise the SSPs in the context of different levels of climate change along multiple Representative Concentration Pathways (RCPs; Moss et al. 2010; van Vuuren et al. 2011). Each RCP reaches a specific level of anthropogenic radiative forcing by 2100, for example, RCP6.0 reaches a level of +6.0 W m^{-2} and RCP1.9 reaches +1.9 W m^{-2}.

Land-use is of great importance in these pathways because of its potential to contribute to continued CO_2 emissions, and capacity to remove CO_2 from the atmosphere. The pace of land-use change described in these pathways over the twenty-first century is much more rapid than that has been seen historically. The median annual global carbon emission from land-use change across the SSPs is 0.8 PgC (3 PgCO$_2$) in 2030, 0.5 PgC (1.9 PgCO$_2$) in 2050 and −0.2 PgC (−0.7 PgCO$_2$) in 2100 (Jia et al. 2019).

In the resulting matrix of SSP and RCP combinations (O'Neill et al. 2016), the change in the global forested area in 2100 (relative to 2010) varies by over 2 billion ha (Fig. 3), equivalent to half of the present-day forested land area. The greatest increases in the forested area are seen in pathways that follow SSP1, which includes strong regulation of the land-sector, increased agricultural productivity, low food waste, and a shift towards lower meat consumption. Accordingly, SSP1 scenarios see reductions in the amount of land used for pasture, and cropland in some realisations (Fig. 3), which allows for forests to regenerate naturally on abandoned land and deliberate afforestation. In the pathways with the most stringent climate target (i.e. RCP1.9) afforestation is a sink of −0.6 PgC (−2.4 $PgCO_2$) per year by 2100 (median value across all SSPs in five IAMs). However, the warming biophysical impacts (i.e. decreased albedo) of forest expansion under some pathways (i.e. RCP4.5) may outweigh the cooling induced by carbon uptake (Davies-Barnard et al. 2014).

The area used for cropland includes both food and energy crops, here the decline in demand for food crops in most SSP1 scenarios is offset by an

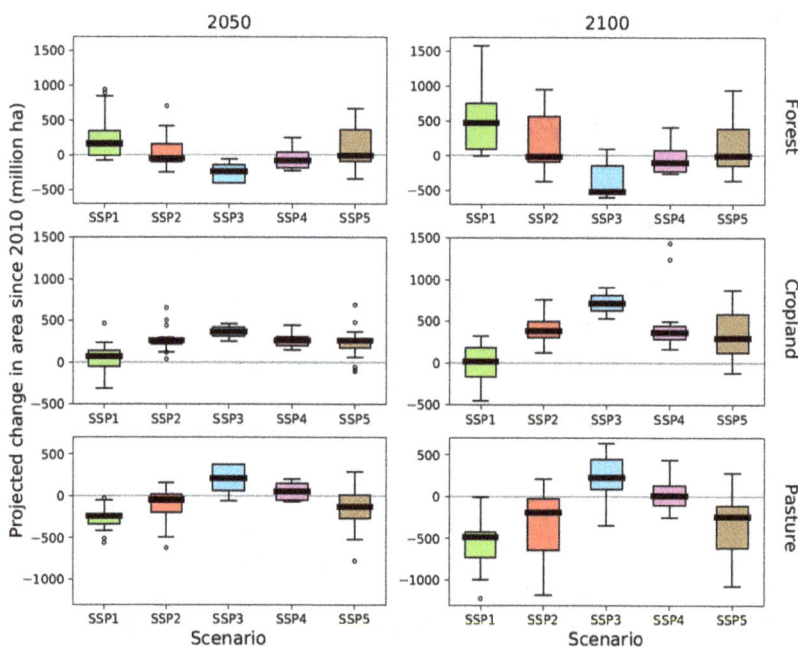

Figure 3 Projected change in land area covered by forest (top panels), cropland (central panels) and pasture (lower panels) in 2050 (left) and 2100 (right), relative to the year 2010, under five Shared Socioeconomic Pathways (Riahi et al. 2017). Each SSP is realised with five integrated assessment models (AIM, GCAM, IMAGE, MESSAGE-GLOBIOM and REMIND/MAgPIE; as described in Popp et al. 2017) under five representative concentration pathways (van Vuuren et al. 2011), data obtained from the IAMC Scenario Explorer (Huppmann et al. 2018).

increased demand for biofuel crop growth. In the pathways with the most stringent climate target, BECCS is a sink of −4 PgC (−14.9 $PgCO_2$) per year by 2100. The greatest increases in cropland and pasture, coupled with a decline in the forested area, are seen in SSP3 scenarios (Fig. 3), where deforestation occurs due to limited regulation and continued competition for resources.

Rather than being predictions, the RCPs and SSPs explore possible future scenarios and are being used by the global climate science community during the Coupled Model Intercomparison Project Phase 6 (CMIP6) to determine the impacts on climate, beyond carbon emissions, associated with different trajectories for the coming century.

6 Future trends and conclusion

Agriculture is currently the main driver of global forest loss, which occurs mainly in the tropics, with land cleared predominantly to grow crops or raise livestock (Gibbs et al. 2010). The specific drivers of land-use change vary regionally and temporally, however, with livestock and soybean growth currently dominating in South America and oil palm growth dominating in Southeast Asia.

Although reconstructing historical land-use change is challenging (Pongratz et al. 2009; Kaplan et al. 2011), it is estimated to have resulted in the emission of around 200 PgC (730 $PgCO_2$) since the year 1750 with current annual emissions estimated at 1.5 ± 0.7 PgC (or 5.5 $PgCO_2$, approximately 14% of total anthropogenic emissions; Le Quéré et al. 2018). Aside from the direct emission of CO_2, the process of land-use change has substantial impacts on the fluxes of energy, moisture and volatile gases between the land surface and the atmosphere.

Computer simulations can be used to assess the overall impact of a particular land-use change; modelling studies find that the overall impact of forests and land-use on climate is latitude dependent. Forests at high northern latitudes exert an overall warming impact on climate due to their low albedo (Betts 2000; Davin and de Noblet-Ducoudré 2010), whereas forests in tropical latitudes sequester huge quantities of carbon, giving them an overall cooling impact on the climate (Bala et al. 2007). The climate impacts of forests at temperate latitudes are less clear (Claussen et al. 2001; Swann et al. 2012), but recent observational studies indicate that temperate forests may have a stronger cooling impact on climate than model simulations have previously suggested.

When combined, the agriculture and land-use sector contribute around one-quarter of current anthropogenic GHG emissions, highlighting the crucial role that land management practices must play in climate change mitigation over the coming decades. Meeting the commitments outlined in the Paris Agreement on Climate, to 'hold the increase in the global average temperature to well below 2°C above pre-industrial levels' and to 'pursue efforts to limit the temperature increase to 1.5°C', is likely to require global CO_2 emissions to reach net zero by around 2050. Achieving net zero will necessitate a complete, or almost complete, elimination of all GHG emissions from agriculture and the land sector, with any remaining emissions being balanced by processes that remove carbon dioxide from the atmosphere.

In addition to reducing direct GHG emissions from agriculture, reducing the amount of land used for food production could enable the natural regeneration, or deliberate replanting, of forests. Scenarios designed to allow the climate science community to explore the impacts of different global futures indicate that a very wide range of land-use trajectories are possible, depending on levels of regulation and cooperation between regions. By 2100, global forested land varies by around 2 billion ha between different realisations of five SSPs combined with RCPs. Current and future research is exploring the impacts of such substantial land-use changes using fully coupled Earth System Models. These models will enable scientists to go beyond quantifying carbon emissions and diagnose the impacts of land-cover change on surface reflectivity, evapotranspiration and the composition of the atmosphere.

7 Where to look for further information

- 'Ecological Climatology' by Gordon Bonan (2016) provides an excellent overview of interactions between the land-surface, atmosphere and climate.
- Global Forest Watch provides up-to-date information on rates of land-cover change around the world: https://www.globalforestwatch.org/.
- Intergovernmental Panel on Climate Change (IPCC) published a Special Report on Climate Change and Land (2019) to collate the latest scientific evidence on the role of LULCC in climate change: https://www.ipcc.ch/report/srccl/. Chapter 2 in particular discusses the links between land-use and climate (Jia et al. 2019).

8 References

Achard, F., Eva, H. D., Stibig, H. J., Mayaux, P., Gallego, J., Richards, T. and Malingreau, J. P. 2002. Determination of deforestation rates of the world's humid tropical forests. *Science* 297(5583), 999–1002. doi:10.1126/science.1070656.

Amin, H. S., Russo, R. S., Sive, B., Richard Hoebeke, E., Dodson, C., McCubbin, I. B., Gannet Hallar, A. and Huff Hartz, K. E. 2013. Monoterpene emissions from bark beetle infested *Engelmann spruce* trees. *Atmospheric Environment* 72(0), 130–3. doi:10.1016/j.atmosenv.2013.02.025.

Angelsen, A. and Wetrtz-Kanounnikoff, S. 2008. What are the key design issues for REDD and the criteria for assessing options? In: Angelsen, A. (Ed.), *Moving Ahead with REDD*. CIFOR, pp. 11–22.

Arora, V. K. and Montenegro, A. 2011. Small temperature benefits provided by realistic afforestation efforts. *Nature Geoscience* 4(8), 514–8. doi:10.1038/ngeo1182.

Ashworth, K., Folberth, G., Hewitt, C. N. and Wild, O. 2012. Impacts of near-future cultivation of biofuel feedstocks on atmospheric composition and local air quality. *Atmospheric Chemistry and Physics* 12(2), 919–39. doi:10.5194/acp-12-919-2012.

Asner, G. P. 2009. Tropical forest carbon assessment: integrating satellite and airborne mapping approaches. *Environmental Research Letters* 4(3). doi:10.1088/1748-9326/4/3/034009.

Baccini, A., Goetz, S. J., Walker, W. S., Laporte, N. T., Sun, M., Sulla-Menashe, D., Hackler, J., Beck, P. S. A., Dubayah, R., Friedl, M. A., Samanta, S. and Houghton, R. A. 2012. Estimated carbon dioxide emissions from tropical deforestation improved by carbon-density maps. *Nature Climate Change* 2(3), 182–5. doi:10.1038/nclimate1354.

Bajzelj, B., Richards, K. S., Allwood, J. M., Smith, P., Dennis, J. S., Curmi, E. and Gilligan, C. A. 2014. Importance of food-demand management for climate mitigation. *Nature Climate Change* 4(10), 924–9. doi:10.1038/nclimate2353.

Bala, G., Caldeira, K., Wickett, M., Phillips, T. J., Lobell, D. B., Delire, C. and Mirin, A. 2007. Combined climate and carbon-cycle effects of large-scale deforestation. *Proceedings of the National Academy of Sciences of the United States of America* 104(16), 6550–5. doi:10.1073/pnas.0608998104.

Bastin, J. F., Finegold, Y., Garcia, C., Mollicone, D., Rezende, M., Routh, D., Zohner, C. M. and Crowther, T. W. 2019. The global tree restoration potential. *Science* 365(6448), 76–9. doi:10.1126/science.aax0848.

Benton, T. G., Bailey, R., Froggatt, A., King, R., Lee, B. and Wellesley, L. 2018. Designing sustainable landuse in a 1.5°C world: the complexities of projecting multiple ecosystem services from land. *Current Opinion in Environmental Sustainability* 31, 88–95. doi:10.1016/j.cosust.2018.01.011.

Betts, R. A. 2000. Offset of the potential carbon sink from boreal forestation by decreases in surface albedo. *Nature* 408(6809), 187–90. doi:10.1038/35041545.

Betts, A. K. and Ball, J. H. 1997. Albedo over the boreal forest. *Journal of Geophysical Research: Atmospheres* 102(D24), 28901–9. doi:10.1029/96JD03876.

Bonan, G. B. 2008. Forests and climate change: forcings, feedbacks, and the climate benefits of forests. *Science* 320(5882), 1444–9. doi:10.1126/science.1155121.

Bonan, G. B., Pollard, D. and Thompson, S. L. 1992. Effects of boreal forest vegetation on global climate. *Nature* 359(6397), 716–8. doi:10.1038/359716a0.

Boserup, E. 1965. *The Conditions of Agricultural Growth: The Economics of Agrarian Change Under Population Pressure*. G. Allen & Unwin, London, UK.

Bounoua, L., DeFries, R., Collatz, G. J., Sellers, P. and Khan, H. 2002. Effects of land cover conversion on surface climate. *Climatic Change* 52(1–2), 29–64.

Bowen, I. S. 1926. The ratio of heat losses by conduction and by evaporation from any water surface. *Physical Review* 27(6), 779–87. doi:10.1103/PhysRev.27.779.

Busch, J., Ferretti-Gallon, K., Engelmann, J., Wright, M., Austin, K. G., Stolle, F., Turubanova, S., Potapov, P. V., Margono, B., Hansen, M. C. and Baccini, A. 2015. Reductions in emissions from deforestation from Indonesia's moratorium on new oil palm, timber, and logging concessions. *Proceedings of the National Academy of Sciences of the United States of America* 112(5), 1328–33. doi:10.1073/pnas.1412514112.

Canadell, J., Jackson, R. B., Ehleringer, J. B., Mooney, H. A., Sala, O. E. and Schulze, E. D. 1996. Maximum rooting depth of vegetation types at the global scale. *Oecologia* 108(4), 583–95. doi:10.1007/BF00329030.

Carvalho, G. O., Nepstad, D., McGrath, D., del Carmen Vera Diaz, M., Santilli, M. and Barros, A. C. 2002. Frontier expansion in the Amazon: balancing development and sustainability. *Environment: Science and Policy for Sustainable Development* 44(3), 34-44. doi:10.1080/00139150209605606.

Ciais, P., Sabine, C., Bala, G., Bopp, L., Brovkin, V., Canadell, J., Chhabra, A., DeFries, R., Galloway, J., Heimann, M., Jones, C., Le Quere, C., Myneni, R. B., Piao, S. and Thornton, P. 2013. Carbon and other biogeochemical cycles. In: Stocker, T. F., Qin, D., Plattner, G.-K., Tignor, M., Allen, S. K., Boschung, J., Nauels, A., Xia, Y., Bex, V. and Midgley, P. M. (Eds), *Climate Change 2013: The Physical Science Basis. Contribution of Working Group I to the Fifth Assessment Report of the Intergovernmental Panel on Climate Change*. Cambridge University Press, Cambridge, UK and New York, NY.

Claussen, M., Brovkin, V. and Ganopolski, A. 2001. Biogeophysical versus biogeochemical feedbacks of large-scale land cover change. *Geophysical Research Letters* 28(6), 1011-4. doi:10.1029/2000GL012471.

Davies-Barnard, T., Valdes, P. J., Singarayer, J. S., Pacifico, F. M. and Jones, C. D. 2014. Full effects of land use change in the representative concentration pathways. *Environmental Research Letters* 9(11). doi:10.1088/1748-9326/9/11/114014.

Davin, E. L. and de Noblet-Ducoudré, N. 2010. Climatic impact of global-scale deforestation: radiative versus nonradiative processes. *Journal of Climate* 23(1), 97-112. doi:10.1175/2009JCLI3102.1.

Defries, R. S., Hansen, M. C., Townshend, J. R. G., Janetos, A. C. and Loveland, T. R. 2000. A new global 1-km dataset of percentage tree cover derived from remote sensing. *Global Change Biology* 6(2), 247-54. doi:10.1046/j.1365-2486.2000.00296.x.

Dixon, R. K., Solomon, A. M., Brown, S., Houghton, R. A., Trexier, M. C. and Wisniewski, J. 1994. Carbon pools and flux of global forest ecosystems. *Science* 263(5144), 185-90. doi:10.1126/science.263.5144.185.

Fang, J.-Y., Wang, G. G., Liu, G.-H. and Xu, S.-L. 1998. Forest biomass of China: an estimate based on the biomass-volume relationship. *Ecological Applications* 8(4), 1084-91.

Fang, J., Chen, A., Peng, C., Zhao, S. and Ci, L. 2001. Changes in forest biomass carbon storage in China between 1949 and 1998. *Science* 292(5525), 2320-2. doi:10.1126/science.1058629.

FAO. 2012. *State of the World's Forests*. United Nations, Rome, Italy.

FAO. 2015. *Global Forest Resources Assessment 2015: How Are the World's Forests Changing?* United Nations, Rome, Italy.

FAO. 2018. *The State of the World's Forests 2018 - Forest Pathways to Sustainable Development*. United Nations, Rome, Italy.

Fearnside, P. M. 2001. Soybean cultivation as a threat to the environment in Brazil. *Environmental Conservation* 28(1), 23-38. doi:10.1017/S0376892901000030.

Fearnside, P. M. 2005. Deforestation in Brazilian Amazonia: history, rates, and consequences. *Conservation Biology* 19(3), 680-8. doi:10.1111/j.1523-1739.2005.00697.x.

Feddema, J., Oleson, K., Bonan, G., Mearns, L., Washington, W., Meehl, G. and Nychka, D. 2005. A comparison of a GCM response to historical anthropogenic land cover change and model sensitivity to uncertainty in present-day land cover representations. *Climate Dynamics* 25(6), 581-609. doi:10.1007/s00382-005-0038-z.

Food and Agriculture Organisation of the United Nations (FAO). 2009. *How to Feed the World in 2050*. United Nations, Rome, Italy.

Forsell, N., Turkovska, O., Gusti, M., Obersteiner, M., Elzen, M. D. and Havlik, P. 2016. Assessing the INDCs' land use, land use change, and forest emission projections. *Carbon Balance and Management* 11(1), 26. doi:10.1186/s13021-016-0068-3.

Friedlingstein, P., Allen, M., Canadell, J. G., Peters, G. P. and Seneviratne, S. I. 2019. Comment on "the global tree restoration potential. *Science* 366(6463), eaay8060. doi:10.1126/science.aay8060.

Georgescu, M., Lobell, D. B. and Field, C. B. 2011. Direct climate effects of perennial bioenergy crops in the United States. *Proceedings of the National Academy of Sciences of the United States of America* 108(11), 4307–12. doi:10.1073/pnas.1008779108.

Gibbard, S., Caldeira, K., Bala, G., Phillips, T. J. and Wickett, M. 2005. Climate effects of global land cover change. *Geophysical Research Letters* 32(23), L23705. doi:10.1029/2005GL024550.

Gibbs, H. K., Johnston, M., Foley, J. A., Holloway, T., Monfreda, C., Ramankutty, N. and Zaks, D. 2008. Carbon payback times for crop-based biofuel expansion in the tropics: the effects of changing yield and technology. *Environmental Research Letters* 3(3). doi:10.1088/1748-9326/3/3/034001.

Gibbs, H. K., Ruesch, A. S., Achard, F., Clayton, M. K., Holmgren, P., Ramankutty, N. and Foley, J. A. 2010. Tropical forests were the primary sources of new agricultural land in the 1980s and 1990s. *Proceedings of the National Academy of Sciences of the United States of America* 107(38), 16732–7.

Gibbs, H. K., Rausch, L., Munger, J., Schelly, I., Morton, D. C., Noojipady, P., Soares-Filho, B., Barreto, P., Micol, L. and Walker, N. F. 2015. Environment and development. Brazil's soy moratorium. *Science* 347(6220), 377–8. doi:10.1126/science.aaa0181.

Gordon, H., Sengupta, K., Rap, A., Duplissy, J., Frege, C., Williamson, C., Heinritzi, M., Simon, M., Yan, C., Almeida, J., Tröstl, J., Nieminen, T., Ortega, I. K., Wagner, R., Dunne, E. M., Adamov, A., Amorim, A., Bernhammer, A. K., Bianchi, F., Breitenlechner, M., Brilke, S., Chen, X., Craven, J. S., Dias, A., Ehrhart, S., Fischer, L., Flagan, R. C., Franchin, A., Fuchs, C., Guida, R., Hakala, J., Hoyle, C. R., Jokinen, T., Junninen, H., Kangasluoma, J., Kim, J., Kirkby, J., Krapf, M., Kürten, A., Laaksonen, A., Lehtipalo, K., Makhmutov, V., Mathot, S., Molteni, U., Monks, S. A., Onnela, A., Peräkylä, O., Piel, F., Petäjä, T., Praplan, A. P., Pringle, K. J., Richards, N. A., Rissanen, M. P., Rondo, L., Sarnela, N., Schobesberger, S., Scott, C. E., Seinfeld, J. H., Sharma, S., Sipilä, M., Steiner, G., Stozhkov, Y., Stratmann, F., Tomé, A., Virtanen, A., Vogel, A. L., Wagner, A. C., Wagner, P. E., Weingartner, E., Wimmer, D., Winkler, P. M., Ye, P., Zhang, X., Hansel, A., Dommen, J., Donahue, N. M., Worsnop, D. R., Baltensperger, U., Kulmala, M., Curtius, J. and Carslaw, K. S. 2016. Reduced anthropogenic aerosol radiative forcing caused by biogenic new particle formation. *Proceedings of the National Academy of Sciences of the United States of America* 113(43), 12053–8. doi:10.1073/pnas.1602360113.

Grainger, A. 1993. The causes of deforestation. In: Grainger, A. (Ed.), *Controlling Tropical Deforestation*. Earthscan, London, UK, pp. 49–68.

Grassi, G., House, J., Dentener, F., Federici, S., den Elzen, M. and Penman, J. 2017. The key role of forests in meeting climate targets requires science for credible mitigation. *Nature Climate Change* 7(3), 220–6. doi:10.1038/nclimate3227.

Guenther, A. B., Monson, R. K. and Fall, R. 1991. Isoprene and monoterpene emission rate variability: observations with eucalyptus and emission rate algorithm development. *Journal of Geophysical Research* 96(D6), 10799–808. doi:10.1029/91JD00960.

Guenther, A. B., Jiang, X., Heald, C. L., Sakulyanontvittaya, T., Duhl, T., Emmons, L. K. and Wang, X. 2012. The Model of Emissions of Gases and Aerosols from Nature version 2.1 (MEGAN2.1): an extended and updated framework for modeling biogenic emissions. *Geoscientific Model Development* 5(6), 1471-92. doi:10.5194/gmd-5-1471-2012.

Gullison, R. E., Frumhoff, P. C., Canadell, J. G., Field, C. B., Nepstad, D. C., Hayhoe, K., Avissar, R., Curran, L. M., Friedlingstein, P., Jones, C. D. and Nobre, C. 2007. Tropical forests and climate policy. *Science* 316(5827), 985-6. doi:10.1126/science.1136163.

Gumpenberger, M., Vohland, K., Heyder, U., Poulter, B., Macey, K., Rammig, A., Popp, A. and Cramer, W. 2010. Predicting pan-tropical climate change induced forest stock gains and losses – implications for REDD. *Environmental Research Letters* 5(1), 014013. doi:10.1088/1748-9326/5/1/014013.

Hallgren, W., Schlosser, C. A., Monier, E., Kicklighter, D., Sokolov, A. and Melillo, J. 2013. Climate impacts of a large-scale biofuels expansion. *Geophysical Research Letters* 40(8), 1624-30. doi:10.1002/grl.50352.

Hansen, M. C., DeFries, R. S., Townshend, J. R. G., Carroll, M., Dimiceli, C. and Sohlberg, R. A. 2003. Global percent tree cover at a spatial resolution of 500 meters: first results of the MODIS vegetation continuous fields algorithm. *Earth Interactions* 7(10), 1-15. doi:10.1175/1087-3562(2003)007<0001:GPTCAA>2.0.CO;2.

Hansen, M. C., Potapov, P. V., Moore, R., Hancher, M., Turubanova, S. A., Tyukavina, A., Thau, D., Stehman, S. V., Goetz, S. J., Loveland, T. R., Kommareddy, A., Egorov, A., Chini, L., Justice, C. O. and Townshend, J. R. 2013. High-resolution global maps of 21st-century forest cover change. *Science* 342(6160), 850-3. doi:10.1126/science.1244693.

Harper, A. B., Powell, T., Cox, P. M., House, J., Huntingford, C., Lenton, T. M., Sitch, S., Burke, E., Chadburn, S. E., Collins, W. J., Comyn-Platt, E., Daioglou, V., Doelman, J. C., Hayman, G., Robertson, E., van Vuuren, D., Wiltshire, A., Webber, C. P., Bastos, A., Boysen, L., Ciais, P., Devaraju, N., Jain, A. K., Krause, A., Poulter, B. and Shu, S. 2018. Land-use emissions play a critical role in land-based mitigation for Paris climate targets. *Nature Communications* 9(1), 2938. doi:10.1038/s41467-018-05340-z.

Hollinger, D. Y., Ollinger, S. V., Richardson, A. D., Meyers, T. P., Dail, D. B., Martin, M. E., Scott, N. A., Arkebauer, T. J., Baldocchi, D. D., Clark, K. L., Curtis, P. S., Davis, K. J., Desai, A. R., Dragoni, D., Goulden, M. L., Gu, L., Katul, G. G., Pallardy, S. G., Paw U, K. T., Schmid, H. P., Stoy, P. C., Suyker, A. E. and Verma, S. B. 2010. Albedo estimates for land surface models and support for a new paradigm based on foliage nitrogen concentration. *Global Change Biology* 16(2), 696-710. doi:10.1111/j.1365-2486.2009.02028.x.

Houghton, R. A. 2003. Revised estimates of the annual net flux of carbon to the atmosphere from changes in land use and land management 1850-2000. *Tellus B* 55(2), 378-90. doi:10.1034/j.1600-0889.2003.01450.x.

Huppmann, D., Kriegler, E., Krey, V., Riahi, K., Rogelj, J., Calvin, K., Humpenoeder, F., Popp, A., Rose, S. K., Weyant, J., Bauer, N., Bertram, C., Bosetti, V., Doelman, J., Drouet, L., Emmerling, J., Frank, S., Fujimori, S., Gernaat, D., Grubler, A., Guivarch, C., Haigh, M., Holz, C., Iyer, G., Kato, E., Keramidas, K., Kitous, A., Leblanc, F., Liu, J.-Y., Löffler, K., Luderer, G., Marcucci, A., McCollum, D., Mima, S., Sands, R. D., Sano, F., Strefler, J., Tsutsui, J., Van Vuuren, D., Vrontisi, Z., Wise, M. and Zhang, R. 2018. IAMC 1.5°C scenario explorer and data hosted by IIASA. Integrated Assessment Modeling Consortium & International Institute for Applied Systems Analysis.

Jia, G., Shevliakova, E., Artaxo, P., De Noblet-Ducoudré, N., Houghton, R., House, J., Kitajima, K., Lennard, C., Popp, A., Sirin, A., Sukumar, R. and Verchot, L. 2019. Land-climate interactions. In: Shukla, P. R., Skea, J., Calvo Buendia, E., Masson-Delmotte, V., Pörtner, H.-O., Roberts, D. C., Zhai, P., Slade, R., Connors, S., van Diemen, R., Ferrat, M., Haughey, E., Luz, S., Neogi, S., Pathak, M., Petzold, J., Portugal Pereira, J., Vyas, P., Huntley, E., Kissick, K., Belkacemi, M. and Malley, J. (Eds), *Climate Change and Land: an IPCC Special Report on Climate Change, Desertification, Land Degradation, Sustainable Land Management, Food Security, and Greenhouse Gas Fluxes in Terrestrial Ecosystems* (in press).

Kaplan, J. O., Krumhardt, K. M. and Zimmermann, N. 2009. The prehistoric and preindustrial deforestation of Europe. *Quaternary Science Reviews* 28(27-28), 3016-34. doi:10.1016/j.quascirev.2009.09.028.

Kaplan, J. O., Krumhardt, K. M., Ellis, E. C., Ruddiman, W. F., Lemmen, C. and Goldewijk, K. K. 2011. Holocene carbon emissions as a result of anthropogenic land cover change. *The Holocene* 21(5), 775-91. doi:10.1177/0959683610386983.

Kessler, A. and Baldwin, I. T. 2001. Defensive function of herbivore-induced plant volatile emissions in nature. *Science* 291(5511), 2141-4. doi:10.1126/science.291.5511.2141.

Kirkby, J., Duplissy, J., Sengupta, K., Frege, C., Gordon, H., Williamson, C., Heinritzi, M., Simon, M., Yan, C., Almeida, J., Tröstl, J., Nieminen, T., Ortega, I. K., Wagner, R., Adamov, A., Amorim, A., Bernhammer, A. K., Bianchi, F., Breitenlechner, M., Brilke, S., Chen, X., Craven, J., Dias, A., Ehrhart, S., Flagan, R. C., Franchin, A., Fuchs, C., Guida, R., Hakala, J., Hoyle, C. R., Jokinen, T., Junninen, H., Kangasluoma, J., Kim, J., Krapf, M., Kürten, A., Laaksonen, A., Lehtipalo, K., Makhmutov, V., Mathot, S., Molteni, U., Onnela, A., Peräkylä, O., Piel, F., Petäjä, T., Praplan, A. P., Pringle, K., Rap, A., Richards, N. A., Riipinen, I., Rissanen, M. P., Rondo, L., Sarnela, N., Schobesberger, S., Scott, C. E., Seinfeld, J. H., Sipilä, M., Steiner, G., Stozhkov, Y., Stratmann, F., Tomé, A., Virtanen, A., Vogel, A. L., Wagner, A. C., Wagner, P. E., Weingartner, E., Wimmer, D., Winkler, P. M., Ye, P., Zhang, X., Hansel, A., Dommen, J., Donahue, N. M., Worsnop, D. R., Baltensperger, U., Kulmala, M., Carslaw, K. S. and Curtius, J. 2016. Ion-induced nucleation of pure biogenic particles. *Nature* 533(7604), 521-6. doi:10.1038/nature17953.

Klein Goldewijk, K. K. 2001. Estimating global land use change over the past 300 years: the HYDE Database. *Global Biogeochemical Cycles* 15(2), 417-33. doi:10.1029/1999GB001232.

Klein Goldewijk, K., Beusen, A., Van Drecht, G. and De Vos, M. 2011. The HYDE 3.1 spatially explicit database of human-induced global land-use change over the past 12,000 years. *Global Ecology and Biogeography* 20(1), 73-86. doi:10.1111/j.1466-8238.2010.00587.x.

Koh, L. P., Miettinen, J., Liew, S. C. and Ghazoul, J. 2011. Remotely sensed evidence of tropical peatland conversion to oil palm. *Proceedings of the National Academy of Sciences of the United States of America* 108(12), 5127-32. doi:10.1073/pnas.1018776108.

Lean, J. and Warrilow, D. A. 1989. Simulation of the regional climatic impact of Amazon deforestation. *Nature* 342(6248), 411-3. doi:10.1038/342411a0.

Leonard, R. E. and Eschner, A. R. 1968. Albedo of intercepted snow. *Water Resources Research* 4(5), 931-5. doi:10.1029/WR004i005p00931.

Le Quéré, C., Andrew, R. M., Friedlingstein, P., Sitch, S., Hauck, J., Pongratz, J., Pickers, P. A., Korsbakken, J. I., Peters, G. P., Canadell, J. G., Arneth, A., Arora, V. K., Barbero, L.,

Bastos, A., Bopp, L., Chevallier, F., Chini, L. P., Ciais, P., Doney, S. C., Gkritzalis, T., Goll, D. S., Harris, I., Haverd, V., Hoffman, F. M., Hoppema, M., Houghton, R. A., Hurtt, G., Ilyina, T., Jain, A. K., Johannessen, T., Jones, C. D., Kato, E., Keeling, R. F., Goldewijk, K. K., Landschützer, P., Lefèvre, N., Lienert, S., Liu, Z., Lombardozzi, D., Metzl, N., Munro, D. R., Nabel, J. E. M. S., Nakaoka, S., Neill, C., Olsen, A., Ono, T., Patra, P., Peregon, A., Peters, W., Peylin, P., Pfeil, B., Pierrot, D., Poulter, B., Rehder, G., Resplandy, L., Robertson, E., Rocher, M., Rödenbeck, C., Schuster, U., Schwinger, J., Séférian, R., Skjelvan, I., Steinhoff, T., Sutton, A., Tans, P. P., Tian, H., Tilbrook, B., Tubiello, F. N., van der Laan-Luijkx, I. T., van der Werf, G. R., Viovy, N., Walker, A. P., Wiltshire, A. J., Wright, R., Zaehle, S. and Zheng, B. 2018. Global carbon budget 2018. *Earth System Science Data* 10(4), 2141-94. doi:10.5194/essd-10-2141-2018.

Lewis, S. L., Mitchard, E. T. A., Prentice, C., Maslin, M. and Poulter, B. 2019. Comment on "the global tree restoration potential. *Science* 366(6463), eaaz0388. doi:10.1126/science.aaz0388.

Loreto, F. and Velikova, V. 2001. Isoprene produced by leaves protects the photosynthetic apparatus against ozone damage, quenches ozone products, and reduces lipid peroxidation of cellular membranes. *Plant Physiology* 127(4), 1781-7. doi:10.1104/pp.010497.

Malhi, Y., Adu-Bredu, S., Asare, R. A., Lewis, S. L. and Mayaux, P. 2013. African rainforests: past, present and future. *Philosophical Transactions of the Royal Society of London. Series B, Biological Sciences* 368(1625), 20120312. doi:10.1098/rstb.2012.0312.

Mayaux, P., Pekel, J. F., Desclée, B., Donnay, F., Lupi, A., Achard, F., Clerici, M., Bodart, C., Brink, A., Nasi, R. and Belward, A. 2013. State and evolution of the African rainforests between 1990 and 2010. *Philosophical Transactions of the Royal Society of London. Series B, Biological Sciences* 368(1625), 20120300. doi:10.1098/rstb.2012.0300.

Miettinen, J., Shi, C. and Liew, S. C. 2011. Deforestation rates in insular Southeast Asia between 2000 and 2010. *Global Change Biology* 17(7), 2261-70. doi:10.1111/j.1365-2486.2011.02398.x.

Monks, P. S., Archibald, A. T., Colette, A., Cooper, O., Coyle, M., Derwent, R., Fowler, D., Granier, C., Law, K. S., Mills, G. E., Stevenson, D. S., Tarasova, O., Thouret, V., von Schneidemesser, E., Sommariva, R., Wild, O. and Williams, M. L. 2015. Tropospheric ozone and its precursors from the urban to the global scale from air quality to short-lived climate forcer. *Atmospheric Chemistry and Physics* 15(15), 8889-973. doi:10.5194/acp-15-8889-2015.

Monteith, J. L. and Unsworth, M. H. 2008. Microclimatology of Radiation (i). In: *Principles of Environmental Physics*. Academic Press, pp.86-99.

Moss, R. H., Edmonds, J. A., Hibbard, K. A., Manning, M. R., Rose, S. K., van Vuuren, D. P., Carter, T. R., Emori, S., Kainuma, M., Kram, T., Meehl, G. A., Mitchell, J. F., Nakicenovic, N., Riahi, K., Smith, S. J., Stouffer, R. J., Thomson, A. M., Weyant, J. P. and Wilbanks, T. J. 2010. The next generation of scenarios for climate change research and assessment. *Nature* 463(7282), 747-56. doi:10.1038/nature08823.

Muller, C. H. 1966. The role of chemical inhibition (allelopathy) in vegetational composition. *Bulletin of the Torrey Botanical Club* 93(5), 332-51. doi:10.2307/2483447.

Myhre, G., Shindell, D., Breon, F.-M., Collins, W., Fuglestvedt, J., Huang, J., Koch, D., Lamarque, J.-F., Lee, D., Mendoza, B., Nakajima, T., Robock, A., Stephens, G., Takemura, T. and Zhang, H. 2013. Anthropogenic and natural radiative forcing. In: Stocker, T. F., Qin, D., Plattner, G.-K., Tignor, M., Allen, S. K., Boschung, J., Nauels, A., Xia, Y., Bex, V. and Midgley, P. M. (Eds), *Climate Change 2013: The Physical*

Science Basis. Contribution of Working Group I to the Fifth Assessment Report of the Intergovernmental Panel on Climate Change. Cambridge University Press, Cambridge, UK and New York, NY.

Nepstad, D. C., de Carvalho, C. R., Davidson, E. A., Jipp, P. H., Lefebvre, P. A., Negreiros, G. H., da Silva, E. D., Stone, T. A., Trumbore, S. E. and Vieira, S. 1994. The role of deep roots in the hydrological and carbon cycles of Amazonian forests and pastures. *Nature* 372(6507), 666–9. doi:10.1038/372666a0.

Nepstad, D. C., Stickler, C. M. and Almeida, O. T. 2006. Globalization of the Amazon soy and beef industries: opportunities for conservation. *Conservation Biology: the Journal of the Society for Conservation Biology* 20(6), 1595–603. doi:10.1111/j.1523-1739.2006.00510.x.

Oh, H. K., Sakai, T., Jones, M. B. and Longhurst, W. M. 1967. Effect of various essential oils isolated from Douglas fir needles upon sheep and deer rumen microbial activity. *Applied Microbiology* 15(4), 777–84. doi:10.1128/AEM.15.4.777-784.1967.

O'Neill, B. C., Tebaldi, C., van Vuuren, D. P., Eyring, V., Friedlingstein, P., Hurtt, G., Knutti, R., Kriegler, E., Lamarque, J., Lowe, J., Meehl, G. A., Moss, R., Riahi, K. and Sanderson, B. M. 2016. The scenario model intercomparison project (ScenarioMIP) for CMIP6. *Geoscientific Model Development* 9(9), 3461–82. doi:10.5194/gmd-9-3461-2016.

Paasonen, P., Asmi, A., Petäjä, T., Kajos, M. K., Äijälä, M., Junninen, H., Holst, T., Abbatt, J. P. D., Arneth, A., Birmili, W., van der Gon, H. D., Hamed, A., Hoffer, A., Laakso, L., Laaksonen, A., Leaitch, W. R., Plass-Dülmer, C., Pryor, S. C., Räisänen, P., Swietlicki, E., Wiedensohler, A., Worsnop, D. R., Kerminen, V. and Kulmala, M. 2013. Warming-induced increase in aerosol number concentration likely to moderate climate change. *Nature Geoscience* 6(6), 438–42. doi:10.1038/ngeo1800.

Pongratz, J., Reick, C., Raddatz, T. and Claussen, M. 2008. A reconstruction of global agricultural areas and land cover for the last millennium. *Global Biogeochemical Cycles* 22(3), GB3018. doi:10.1029/2007GB003153.

Pongratz, J., Reick, C. H., Raddatz, T. and Claussen, M. 2009. Effects of anthropogenic land cover change on the carbon cycle of the last millennium. *Global Biogeochemical Cycles* 23(4), n/a–. doi:10.1029/2009GB003488.

Pongratz, J., Reick, C. H., Raddatz, T. and Claussen, M. 2010. Biogeophysical versus biogeochemical climate response to historical anthropogenic land cover change. *Geophysical Research Letters* 37(8), L08702. doi:10.1029/2010GL043010.

Pongratz, J., Reick, C. H., Raddatz, T., Caldeira, K. and Claussen, M. 2011. Past land use decisions have increased mitigation potential of reforestation. *Geophysical Research Letters* 38(15), L15701. doi:10.1029/2011GL047848.

Popp, A., Calvin, K., Fujimori, S., Havlik, P., Humpenöder, F., Stehfest, E., Bodirsky, B. L., Dietrich, J. P., Doelmann, J. C., Gusti, M., Hasegawa, T., Kyle, P., Obersteiner, M., Tabeau, A., Takahashi, K., Valin, H., Waldhoff, S., Weindl, I., Wise, M., Kriegler, E., Lotze-Campen, H., Fricko, O., Riahi, K. and Vuuren, D. Pv 2017. Land-use futures in the shared socio-economic pathways. *Global Environmental Change* 42, 331–45. doi:10.1016/j.gloenvcha.2016.10.002.

Prentice, I. C., Farquhar, G. D., Fasham, M. J. R., Goulden, M. L., Heimann, M., Jaramillo, V. J., Kheshgi, H. S., Le Quéré, C., Scholes, R. J. and Wallace, D. W. R. 2001. The carbon cycle and atmospheric carbon dioxide. In: Houghton, J. T., Ding, Y., Griggs, D. J., et al. *Climate Change 2001: the Physical Science Basis. Contribution of Working Group I to the Third Assessment Report of the Intergovernmental Panel on Climate Change.* Cambridge University Press, Cambridge, UK and New York, NY.

Ramankutty, N. and Foley, J. A. 1999. Estimating historical changes in global land cover: croplands from 1700 to 1992. *Global Biogeochemical Cycles* 13(4), 997-1027. doi:10.1029/1999GB900046.

Ramankutty, N., Evan, A. T., Monfreda, C. and Foley, J. A. 2008. Farming the planet: 1. Geographic distribution of global agricultural lands in the year 2000. *Global Biogeochemical Cycles* 22(1). doi:10.1029/2007GB002952.

Rasmann, S., Köllner, T. G., Degenhardt, J., Hiltpold, I., Toepfer, S., Kuhlmann, U., Gershenzon, J. and Turlings, T. C. 2005. Recruitment of entomopathogenic nematodes by insect-damaged maize roots. *Nature* 434(7034), 732-7. doi:10.1038/nature03451.

Rasmussen, R. A. and Went, F. W. 1965. Volatile organic material of plant origin in the atmosphere. *Proceedings of the National Academy of Sciences of the United States of America* 53(1), 215-20. doi:10.1073/pnas.53.1.215.

Riahi, K., van Vuuren, D. P., Kriegler, E., Edmonds, J., O'Neill, B. C., Fujimori, S., Bauer, N., Calvin, K., Dellink, R., Fricko, O., Lutz, W., Popp, A., Cuaresma, J. C., Kc, S., Leimbach, M., Jiang, L., Kram, T., Rao, S., Emmerling, J., Ebi, K., Hasegawa, T., Havlik, P., Humpenöder, F., Da Silva, L. A., Smith, S., Stehfest, E., Bosetti, V., Eom, J., Gernaat, D., Masui, T., Rogelj, J., Strefler, J., Drouet, L., Krey, V., Luderer, G., Harmsen, M., Takahashi, K., Baumstark, L., Doelman, J. C., Kainuma, M., Klimont, Z., Marangoni, G., Lotze-Campen, H., Obersteiner, M., Tabeau, A. and Tavoni, M. 2017. The Shared Socioeconomic Pathways and their energy, land use, and greenhouse gas emissions implications: an overview. *Global Environmental Change* 42, 153-68. doi:10.1016/j.gloenvcha.2016.05.009.

Richardson, A. D., Keenan, T. F., Migliavacca, M., Ryu, Y., Sonnentag, O. and Toomey, M. 2013. Climate change, phenology, and phenological control of vegetation feedbacks to the climate system. *Agricultural and Forest Meteorology* 169, 156-73. doi:10.1016/j.agrformet.2012.09.012.

Robinson, D. A. and Kukla, G. 1985. Maximum surface albedo of seasonally snow-covered lands in the northern hemisphere. *Journal of Climate and Applied Meteorology* 24(5), 402-11. doi:10.1175/1520-0450(1985)024<0402:MSAOSS>2.0.CO;2.

Rogelj, J., Shindell, D., Jiang, K., Fifita, S., Forster, P., Ginzburg, V., Handa, C., Kheshgi, H., Kobayashi, S., Kriegler, E., Mundaca, L., Séférian, R. and Vilariño, M. V. 2018. Mitigation pathways compatible with 1.5°C in the context of sustainable development. In: Masson-Delmotte, V., Zhai, P., Pörtner, H.-O., et al. (Eds), *Global Warming of 1.5°C. An IPCC Special Report on the Impacts of Global Warming of 1.5°C above Pre-industrial Levels and Related Global Greenhouse Gas Emission Pathways, in the Context of Strengthening the Global Response to the Threat of Climate Change, Sustainable Development, and Efforts to Eradicate Poverty* (in press).

Ruddiman, W. F. J. C. C. 2003. The anthropogenic greenhouse era began thousands of years ago. *Climatic Change* 61(3), 261-93. doi:10.1023/B:CLIM.0000004577.17928.fa.

Rudorff, B. F. T., Adami, M., Aguiar, D. A., Moreira, M. A., Mello, M. P., Fabiani, L., Amaral, D. F. and Pires, B. M. 2011. The soy moratorium in the Amazon biome monitored by remote sensing images. *Remote Sensing* 3(1), 185-202. doi:10.3390/rs3010185.

Saatchi, S., Marlier, M., Chazdon, R. L., Clark, D. B. and Russell, A. E. 2011a. Impact of spatial variability of tropical forest structure on radar estimation of aboveground biomass. *Remote Sensing of Environment* 115(11), 2836-49. doi:10.1016/j.rse.2010.07.015.

Saatchi, S. S., Harris, N. L., Brown, S., Lefsky, M., Mitchard, E. T., Salas, W., Zutta, B. R., Buermann, W., Lewis, S. L., Hagen, S., Petrova, S., White, L., Silman, M. and Morel, A. 2011b. Benchmark map of forest carbon stocks in tropical regions across three continents. *Proceedings of the National Academy of Sciences of the United States of America* 108(24), 9899–904. doi:10.1073/pnas.1019576108.

Sanadze, G. A. and Kursanov, A. L. 1966. On certain conditions of the evolution of the diene C_5H_8 from poplar leaves. *Soviet Plant Physiology* 13, 184–9.

Schoneveld, G. C., Ekowati, D., Andrianto, A. and van der Haar, S. 2019. Modeling peat- and forestland conversion by oil palm smallholders in Indonesian Borneo. *Environmental Research Letters* 14(1), (014006). doi:10.1088/1748-9326/aaf044.

Scott, C. E., Rap, A., Spracklen, D. V., Forster, P. M., Carslaw, K. S., Mann, G. W., Pringle, K. J., Kivekäs, N., Kulmala, M., Lihavainen, H. and Tunved, P. 2014. The direct and indirect radiative effects of biogenic secondary organic aerosol. *Atmospheric Chemistry and Physics* 14(1), 447–70. doi:10.5194/acp-14-447-2014.

Scott, C. E., Arnold, S. R., Monks, S. A., Asmi, A., Paasonen, P. and Spracklen, D. V. 2018a. Substantial large-scale feedbacks between natural aerosols and climate. *Nature Geoscience* 11(1), 44–8. doi:10.1038/s41561-017-0020-5.

Scott, C. E., Monks, S. A., Spracklen, D. V., Arnold, S. R., Forster, P. M., Rap, A., Äijälä, M., Artaxo, P., Carslaw, K. S., Chipperfield, M. P., Ehn, M., Gilardoni, S., Heikkinen, L., Kulmala, M., Petäjä, T., Reddington, C. L. S., Rizzo, L. V., Swietlicki, E., Vignati, E. and Wilson, C. 2018b. Impact on short-lived climate forcers increases projected warming due to deforestation. *Nature Communications* 9(1), 157. doi:10.1038/s41467-017-02412-4.

Smith, P., Davis, S. J., Creutzig, F., Fuss, S., Minx, J., Gabrielle, B., Kato, E., Jackson, R. B., Cowie, A., Kriegler, E., van Vuuren, D. P., Rogelj, J., Ciais, P., Milne, J., Canadell, J. G., McCollum, D., Peters, G., Andrew, R., Krey, V., Shrestha, G., Friedlingstein, P., Gasser, T., Grübler, A., Heidug, W. K., Jonas, M., Jones, C. D., Kraxner, F., Littleton, E., Lowe, J., Moreira, J. R., Nakicenovic, N., Obersteiner, M., Patwardhan, A., Rogner, M., Rubin, E., Sharifi, A., Torvanger, A., Yamagata, Y., Edmonds, J. and Yongsung, C. 2016. Biophysical and economic limits to negative CO_2 emissions. *Nature Climate Change* 6(1), 42–50. doi:10.1038/nclimate2870.

Snyder, P. K., Delire, C. and Foley, J. A. 2004. Evaluating the influence of different vegetation biomes on the global climate. *Climate Dynamics* 23(3-4), 279–302. doi:10.1007/s00382-004-0430-0.

Spracklen, D. V., Bonn, B. and Carslaw, K. S. 2008. Boreal forests, aerosols and the impacts on clouds and climate. *Philosophical Transactions. Series A, Mathematical, Physical, and Engineering Sciences* 366(1885), 4613–26. doi:10.1098/rsta.2008.0201.

Spracklen, D. V., Arnold, S. R. and Taylor, C. M. 2012. Observations of increased tropical rainfall preceded by air passage over forests. *Nature* 489(7415), 282–5. doi:10.1038/nature11390.

Spracklen, D. V., Baker, J. C. A., Garcia-Carreras, L. and Marsham, J. H. 2018. The effects of tropical vegetation on rainfall. *Annual Review of Environment and Resources* 43(1), 193–218. doi:10.1146/annurev-environ-102017-030136.

Springmann, M., Godfray, H. C., Rayner, M. and Scarborough, P. 2016. Analysis and valuation of the health and climate change cobenefits of dietary change. *Proceedings of the National Academy of Sciences of the United States of America* 113(15), 4146–51. doi:10.1073/pnas.1523119113.

Stehfest, E., Bouwman, L., van Vuuren, D. P., den Elzen, M. G. J., Eickhout, B. and Kabat, P. 2009. Climate benefits of changing diet. *Climatic Change* 95(1-2), 83-102. doi:10.1007/s10584-008-9534-6.

Swann, A. L. S., Fung, I. Y. and Chiang, J. C. 2012. Mid-latitude afforestation shifts general circulation and tropical precipitation. *Proceedings of the National Academy of Sciences of the United States of America* 109(3), 712-6. doi:10.1073/pnas.1116706108.

Teuling, A. J., Taylor, C. M., Meirink, J. F., Melsen, L. A., Miralles, D. G., van Heerwaarden, C. C., Vautard, R., Stegehuis, A. I., Nabuurs, G. J. and de Arellano, J. V. 2017. Observational evidence for cloud cover enhancement over Western European forests. *Nature Communications* 8, 14065. doi:10.1038/ncomms14065.

Thomas, G. and Rowntree, P. R. 1992. The boreal forests and climate. *Quarterly Journal of the Royal Meteorological Society* 118(505), 469-97. doi:10.1002/qj.49711850505.

Unger, N. 2014. Human land-use-driven reduction of forest volatiles cools global climate. *Nature Climate Change* 4(10), 907-10. doi:10.1038/nclimate2347.

USDA-FAS. 2019. *Oilseeds: World Markets and Trade*. Circular Series FOP July 2019. US Department of Agriculture - Foreign Agriculture Service, Washington DC.

van der Werf, G. R., Randerson, J. T., Giglio, L., Collatz, G. J., Mu, M., Kasibhatla, P. S., Morton, D. C., DeFries, R. S., Jin, Y. and van Leeuwen, T. T. 2010. Global fire emissions and the contribution of deforestation, savanna, forest, agricultural, and peat fires (1997-2009). *Atmospheric Chemistry and Physics* 10(23), 11707-35.

van Vuuren, D. P., Edmonds, J., Kainuma, M., Riahi, K., Thomson, A., Hibbard, K., Hurtt, G. C., Kram, T., Krey, V., Lamarque, J., Masui, T., Meinshausen, M., Nakicenovic, N., Smith, S. J. and Rose, S. K. 2011. The representative concentration pathways: an overview. *Climatic Change* 109(1-2), 5-31. doi:10.1007/s10584-011-0148-z.

Veldman, J. W., Aleman, J. C., Alvarado, S. T., Anderson, T. M., Archibald, S., Bond, W. J., Boutton, T. W., Buchmann, N., Buisson, E., Canadell, J. G., Dechoum, M. S., Diaz-Toribio, M. H., Durigan, G., Ewel, J. J., Fernandes, G. W., Fidelis, A., Fleischman, F., Good, S. P., Griffith, D. M., Hermann, J. M., Hoffmann, W. A., Le Stradic, S., Lehmann, C. E. R., Mahy, G., Nerlekar, A. N., Nippert, J. B., Noss, R. F., Osborne, C. P., Overbeck, G. E., Parr, C. L., Pausas, J. G., Pennington, R. T., Perring, M. P., Putz, F. E., Ratnam, J., Sankaran, M., Schmidt, I. B., Schmitt, C. B., Silveira, F. A. O., Staver, A. C., Stevens, N., Still, C. J., Strömberg, C. A. E., Temperton, V. M., Varner, J. M. and Zaloumis, N. P. 2019. Comment on "the global tree restoration potential". *Science* 366(6463), eaay7976. doi:10.1126/science.aay7976.

Vickers, C. E., Gershenzon, J., Lerdau, M. T. and Loreto, F. 2009. A unified mechanism of action for volatile isoprenoids in plant abiotic stress. *Nature Chemical Biology* 5(5), 283-91. doi:10.1038/nchembio.158.

Wang, S., Chen, J. M., Ju, W. M., Feng, X., Chen, M., Chen, P. and Yu, G. 2007. Carbon sinks and sources in China's forests during 1901-2001. *Journal of Environmental Management* 85(3), 524-37. doi:10.1016/j.jenvman.2006.09.019.

Wang, J., Chagnon, F. J., Williams, E. R., Betts, A. K., Renno, N. O., Machado, L. A., Bisht, G., Knox, R. and Bras, R. L. 2009. Impact of deforestation in the Amazon Basin on cloud climatology. *Proceedings of the National Academy of Sciences of the United States of America* 106(10), 3670-4. doi:10.1073/pnas.0810156106.

Went, F. W. 1960. Organic matter in the atmosphere, and its possible relation to petroleum formation. *Proceedings of the National Academy of Sciences of the United States of America* 46(2), 212-21. doi:10.1073/pnas.46.2.212.

Williams, M. 2003. *Deforesting the Earth*. The University of Chigaco Press.

Zhu, J., Penner, J. E., Lin, G., Zhou, C., Xu, L. and Zhuang, B. 2017. Mechanism of SOA formation determines magnitude of radiative effects. *Proceedings of the National Academy of Sciences of the United States of America* 114(48), 12685-90. doi:10.1073/pnas.1712273114.

Zhu, J., Penner, J. E., Yu, F., Sillman, S., Andreae, M. O. and Coe, H. 2019. Decrease in radiative forcing by organic aerosol nucleation, climate, and land use change. *Nature Communications* 10(1), 423. doi:10.1038/s41467-019-08407-7.

Chapter 2

Understanding how land-use management affects soil microbiomes

Lucas William Mendes, Thierry Alexandre Pellegrinetti and Alexandre Pedrinho, Center for Nuclear Energy in Agriculture, University of São Paulo, Brazil; and Dennis Goss-Souza, Federal Institute of Paraná, Brazil

1 Introduction

Soil is a crucial biological ecosystem, serving as a habitat for billions of microorganisms and supporting rich biodiversity (Bach et al., 2020). The soil microbiome, a complex network of microorganisms inhabiting the soil, plays a pivotal role in sustaining the health and functionality of terrestrial ecosystems. This microbiome encompasses bacteria, fungi, archaea, protozoa, and viruses, collectively contributing to crucial ecosystem services. They participate in nutrient cycling, decomposition of organic matter, soil formation, and even the regulation of greenhouse gas (GHG) emissions (Mendes et al., 2017). The diversity and functions of these microbial communities are intricately linked to soil health and fertility. Land-use change, such as deforestation, urbanization, agricultural intensification, and other human activities, significantly impacts the soil microbiome, leading to alterations in its composition and functionality (Averill et al., 2022).

For instance, deforestation disrupts the delicate balance of microbial communities by altering environmental conditions and reducing organic matter inputs. Likewise, intensive agricultural practices, which involve pesticides, fertilizers, and tillage methods, can significantly impact the diversity and functions

http://dx.doi.org/10.19103/AS.2024.0136.33

of soil microbes. These alterations in the soil microbiome trigger cascading effects on ecosystem functions, agricultural productivity, and even global carbon cycling. Furthermore, disruptions in the soil microbiome resulting from land-use changes can lead to unforeseen consequences, such as increased soil erosion, reduced fertility, and decreased resilience to environmental stressors like droughts or extreme weather events. Understanding these intricate relationships is crucial as they directly influence soil fertility, plant health, and ultimately, the sustainability of ecosystems and human livelihoods.

The Global Land Outlook report (United Nations, 2017) highlights that approximately 25% of the world's land is currently degraded. This degradation often leads to biodiversity loss, soil erosion, and diminished agricultural productivity (Cardinale et al., 2012), affecting the soil's natural productivity and its capacity to support both human and natural systems (Lal, 2020). Hence, studying the soil microbiome in the context of land-use changes becomes imperative for developing sustainable land management practices. This approach aims to preserve soil health, enhance agricultural productivity, mitigate climate change impacts, and safeguard biodiversity in terrestrial ecosystems. This chapter delves into examples illustrating how land-use changes affect soil microbiome composition and functions, focusing on various biomes in Brazil.

2   Effects of land-use change on soil microbiome composition

Land-use changes have profound implications for soil microbial communities, altering their composition, diversity, and functionality. When natural habitats are converted for agricultural, urban, or industrial purposes, these alterations can disrupt the delicate balance within soil ecosystems. Intensive land-use practices like deforestation or agricultural expansion often lead to decreased microbial diversity and altered community structures. For instance, the application of pesticides and chemical fertilizers in agriculture can result in the depletion of certain microbial species while favoring the proliferation of others, thus impacting the overall microbial balance. Additionally, the removal of vegetation cover through land conversion diminishes organic matter inputs to the soil, affecting microbial food sources and disrupting their ecological functions, such as nutrient cycling and decomposition processes.

Furthermore, land-use changes can induce shifts in soil microbial activity and functionality. For example, deforestation along with agricultural practices can lead to increased soil compaction, altering microbial habitats and reducing their ability to carry out essential functions. Changes in land use may also modify soil pH, moisture levels, and nutrient availability, directly influencing the metabolic activities and survival of various microbial communities. Such alterations can affect the breakdown of organic matter and nutrient cycling rates, ultimately impacting the overall soil fertility. Understanding these intricate relationships between land-use changes and soil microbial communities is crucial for devising sustainable land management strategies that aim to preserve soil health and ecosystem functionality. Figure 1 illustrates the results discussed in this section.

2.1 Effects of land-use change on soil microbiome in the Amazon region

The Amazon rainforest holds 40% of the remaining tropical rainforests, serving as a crucial force in conserving biodiversity, regulating biogeochemical cycles, and managing the climate. Considered the planet's largest repository of plant and animal diversity, the Amazon also shelters a myriad of microorganisms crucial to its ecological balance (Rodrigues et al., 2013). However, despite its immense global significance, the Amazon faces an ongoing threat due to agricultural expansion, marking it as the most active frontier of deforestation worldwide. Deforestation in the Brazilian Amazon involves the clearing and transformation of native forests into alternative land uses, predominantly for agriculture and cattle ranching. The rapid expansion of agricultural activities has become the primary force disrupting the Amazon region, exerting profound consequences on its intricate network of soil microorganisms (Mendes et al., 2015a). Consequently, there has been a recent surge in attention directed toward exploring the belowground microbial communities thriving in Amazonian soils.

The pioneering exploration of microbial diversity in Amazon soils traces back to Borneman and Triplett (1997), who utilized the Sanger sequencing technique to examine sequences extracted from both native forest and pasture areas. Their initial findings unveiled an expansive microbial panorama, underscoring stark disparities in composition between forest and pasture ecosystems. *Clostridia*-affiliated sequences were prevalent in forest samples, while *Bacillus* dominated the pasture environments. Subsequent studies in the Brazilian Amazon have significantly enriched our understanding of how alterations in land use affect soil microbial communities (Jesus et al., 2009; Grossman et al., 2010; Pazinato et al., 2010; Navarrete et al., 2011, 2015; Germano et al., 2012; Rodrigues et al., 2013; Taketani et al., 2013; Brossi et al., 2014; Mendes et al., 2014). These investigations underscore the remarkably diverse microbial communities thriving in Amazonian soils and illuminate how land-use changes profoundly alter the structure, diversity, and composition of these microbial assemblages.

Recent studies aimed at understanding the impact of land-use changes on Amazon soils utilized a soil DNA shotgun metagenomics approach to examine

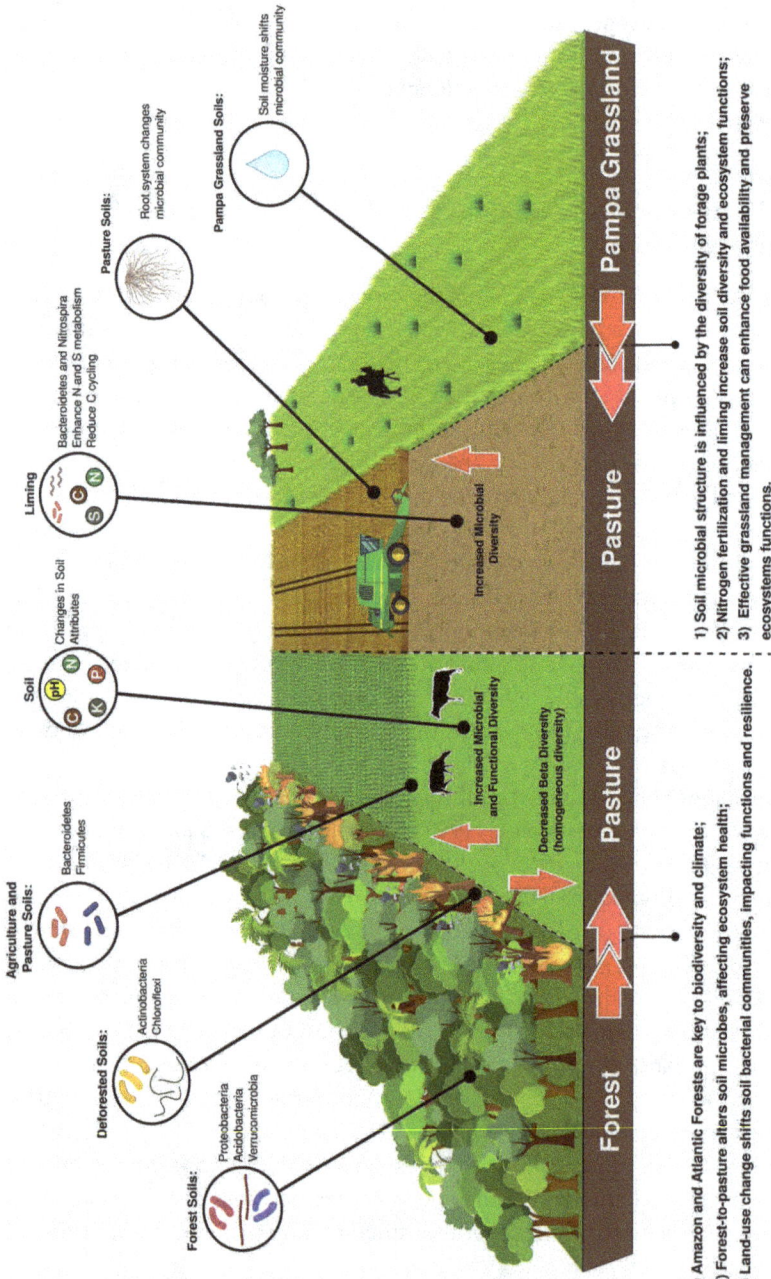

Figure 1 Effects of land-use change on the soil microbiome. The figure illustrates how land-use change affects soil chemical attributes and subsequently influences the soil microbiome.

microbes in various zones: native forest, deforested land converted into soybean fields, and pasture (Mendes et al., 2015a; Pedrinho et al., 2019). These investigations revealed that different land-use systems significantly influenced the types of bacteria present in the soil. Factors such as pH levels and the presence of specific elements, such as carbon, nitrogen, nitrate, and potassium, were closely associated with the variations in bacterial communities. For instance, Proteobacteria, Acidobacteria, and Verrucomicrobia were dominant in forest soils. Conversely, deforested regions showed a higher abundance of Chloroflexi and Actinobacteria, while agriculture and pasture displayed an increased abundance of Bacteroidetes and Firmicutes, respectively. These distinct bacterial populations were linked to their specialized roles within the environment. Proteobacteria, abundant in forest soils, play a crucial role in the global cycling of carbon, nitrogen, and sulfur. Acidobacteria influence the acidity and nutrient properties of Amazon soils, while Verrucomicrobia thrive in forests due to their ability to break down robust carbon compounds found in moist forest soils. Changes in bacterial composition post-deforestation were attributed to heightened soil temperatures and carbon deposition from burning. Consequently, Chloroflexi, thriving in higher temperatures, and Actinobacteria, known for decomposing organic matter, increased in these areas. Bacteroidetes, associated with aiding plant growth and cellulose decomposition, were more abundant in soybean fields. Firmicutes, known for their resilience in harsh conditions, dominated pasture soils, which are typically more challenging environments.

In summary, diverse bacterial communities thrive in different environments based on their unique abilities to adapt to specific conditions and fulfill distinct ecological functions. Thus, the transformation of land use plays a pivotal role in shaping soil microbial communities. Alterations in land use, whether transitioning from native forests to agricultural fields or converting landscapes for different purposes like pasture or cultivation, exert a significant influence on the composition and diversity of soil microbes. The shift from one land use to another triggers a cascade of modifications in soil conditions, leading to specific adaptations among microbial communities to survive and thrive in these altered environments. Consequently, the way land is utilized stands as a crucial determinant in structuring and molding the intricate and diverse microbial communities found within soil ecosystems.

2.2 Effects of land-use change on soil microbiome in the Atlantic Forest

The Brazilian Atlantic Forest ranks as one of the world's richest biodiversity hotspots, hosting 2.7% and 2.1% of global endemic plant and vertebrate species, respectively (Myers et al., 2000). Regrettably, in recent decades, we have witnessed extensive fragmentation and deforestation, resulting in the depletion of the original vegetation to a mere 11.7% (Ribeiro et al., 2009). The conversion of these forests into croplands and pasturelands, constituting 20% and 42% of total human net primary production appropriation, significantly contributes to this loss (Weinzettel et al., 2018). Projections for 2100 suggest a potential 58% decrease in natural vegetative cover across 34 global biodiversity hotspots, with up to 1/3 of habitat loss and 16% of species loss attributed to forest conversion alone (Jantz et al., 2015).

Soil microorganisms, like plants and animals, exhibit significant responses to changes in land use (Lauber et al., 2013; Kaiser et al., 2016; Goss-Souza et al., 2017; Ceola et al., 2021). Studies in Brazilian tropics indicate that converting forests into pasturelands and croplands often results in the loss of bacterial diversity (Jesus et al., 2009; Rodrigues et al., 2013; Mendes et al., 2015b; Goss-Souza et al., 2020) and affects ecosystem services linked to microbial activity (Paula et al., 2014; Meyer et al., 2017; Goss-Souza et al., 2019; Pedrinho et al., 2019). Most research highlights taxa trade-offs, diversity turnover, and shifts in microbial functions due to land-use changes, driven by local abiotic environmental filters such as soil pH and fertility, suggesting a dominance of homogeneous selection (Goss-Souza et al., 2020).

In the Atlantic Forest, limited research, especially in the subtropical region (Southern Brazil), has explored the diversity of soil bacterial communities and the consequences of forest-to-agriculture conversion on both bacterial and arbuscular mycorrhizal (AM) fungi diversities and ecological processes shaping their distribution (Faoro et al., 2010; Goss-Souza et al., 2017; Ceola et al., 2021; Goss-Souza et al., 2022). From a biogeographic perspective, some authors propose that soil type, a historical contingency along with dispersal (an evolutionary contingency), may locally filter taxa distributions more strongly than the influence of land-use change (Ceola et al., 2021; Goss-Souza et al., 2022). These works indicate that soil microbial diversity and niche occupancy are influenced by spatial distance and long-term historical contingencies related to soil origin (soil type and climate). This leads to significant patterns of dispersal limitation and spatial correlations, with stochastic processes outweighing the effect of deterministic selection processes caused by soil historical contingencies and the formation of small geographic islands shaped by soil type and climate. These patterns become particularly evident when evaluating microbial niche specialists, especially at a local scale, with implications for biotic interactions among members of local microbial communities. This emphasizes the importance of expanding our understanding of the microbial component in ecological studies and conservation efforts within the context of the Brazilian Atlantic Forest.

2.3 Effects of land-use change on soil microbiome in grasslands

Over the last century, the escalating demand for primary products such as wood, food, and fiber has precipitated extensive conversion of natural grasslands into cultivated pastures (Sühs et al., 2020). This transition frequently leads to a decrease in the diversity of forage species and consequent impacts on soil microbial diversity (Delory et al., 2019; Yin et al., 2019). Natural and improved natural grasslands often contend with challenges such as overgrazing or exclusion. The heightened demand for forage has instigated the transformation of natural grasslands into cultivated pastures (Zanella et al., 2021). The intensification of soil management has been observed to stimulate the deterministic (niche-based) assembly process of soil microbial communities (Wang et al., 2021). Indeed, soil management enhances microbial functions, potentially owing to niche specialization (Wu et al., 2019). Nevertheless, the impacts of replacing natural grasslands with cultivated pastures on habitat specialization and the resultant assembly of microbial communities remain unknown.

In a recent study (Tomazelli et al., 2023a), the authors demonstrated that the structure of soil microbial communities was influenced by the diversity of forage plants following the conversion of natural grasslands to cultivated pastures. Soil management practices such as nitrogen fertilization and liming increased soil microbial alpha diversity and altered potential ecosystem functions. The study revealed that soil liming was the primary driver of changes in microbial communities by reducing available aluminum levels in the soil. This favored microbial groups related to nitrogen and sulfur metabolism, including 'nitrification,' 'nitrate reduction,' 'nitrogen respiration,' and 'respiration of sulfur compounds,' represented by Bacteroidetes and Nitrospirae. This came with a decrease in potential functions associated with C cycling, such as 'cellulose,' as represented by Acidobacteria.

Another recent study underscored the importance of grassland management and plant diversity in shaping microbial communities in grasslands (Tomazelli et al., 2023b). The authors found that generalists in the grasslands were shaped by niche-based models, while specialists in natural grasslands exhibited more stochastic behavior. In improved pastures, lognormal and preemption models were predominant, suggesting that homogenizing dispersal outweighed variable selection. The study demonstrated that improved pastures are environmentally beneficial, being more deterministic and having a higher proportion of niche-specialist microbes.

In the cultivated pasture soils of southern Brazil, the architecture of root systems, providing high interconnectivity, often results in stochastic behavior in microbial communities, leading to community homogenization in the face of environmental changes (Goss-Souza et al., 2017). In the same region, natural grasslands of the pampa biome exhibited both stochastic and deterministic processes, with soil moisture being the primary deterministic driver of soil microbial community assembly patterns (Lupatini et al., 2019). These results have highlighted that improved management of natural grasslands could serve as an alternative to increase cattle food availability. This approach maintains the endemic diversity of forage plants and soil biological quality while preserving essential ecosystem functions performed by soil microbes.

3   Effects of land-use change on soil microbiome functions

Forest-to-pasture conversion stands as the most prevalent and high-impact type of land-use change in the Brazilian Amazon region (Nascimento et al., 2019; Nunes et al., 2022). Over the years, various studies conducted in this region consistently demonstrated the negative effects of such conversions on vegetation cover (Nunes et al., 2022; Vieira, 2019), soil physicochemical properties (Navarrete et al., 2015; Melo et al., 2017), and soil microorganisms (Goss-Souza et al., 2017; Mendes et al., 2015a; Rodrigues et al., 2013). Recent research has further revealed that forest-to-pasture conversion influences soil microbial functions (Mendes et al., 2015b; Navarrete et al., 2015; Paula et al., 2014; Pedrinho et al., 2019). In this context, we present studies illustrating how land-use change, primarily forest-to-pasture conversion in the Brazilian Amazon region, impacts soil microbial functions, focusing particularly on nitrogen cycling, phosphorus transformation, antibiotic resistance genes (ARGs), and GHG emissions. Figure 2 illustrates the results discussed in this section.

3.1 Effects of land-use change on nitrogen cycling in the Amazon region

The Amazon rainforest plays a crucial role in nitrogen cycling (Pajares and Bohannan, 2016; Pedrinho et al., 2020). Typically, Amazonian soils, along with many other tropical soils, are highly weathered and nutrient-depleted (Fujii et al., 2018). Their fertility heavily relies on the cycling of a thin layer of organic matter associated with a substantial amount of plant litter material (Pajares and Bohannan, 2016). In this context, soil microorganisms play a fundamental role, responsible for both (i) mineralization, which involves the decomposition and/or oxidation of organic matter into forms easily available to plants and/or other soil microorganisms (Elrys et al., 2023), and (ii) assimilative/dissimilative processes, such as nitrogen fixation, nitrification, denitrification, dissimilatory nitrate reduction to ammonium (DNRA), and anammox (Pajares and Bohannan,

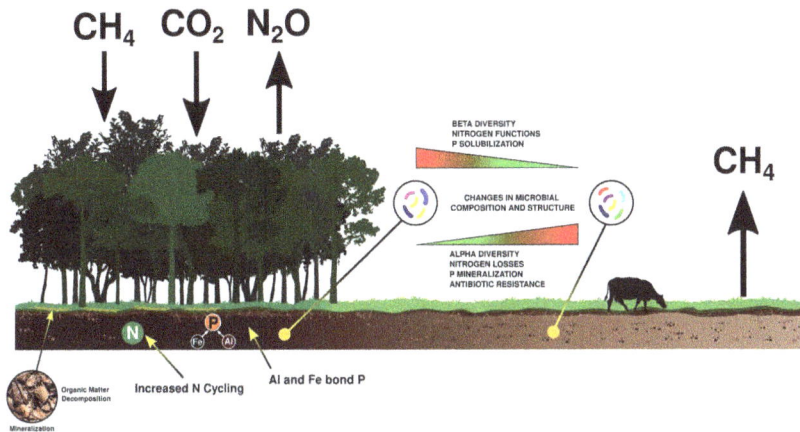

Figure 2 Effects of land-use change on the soil microbiome functions. Native forests act as a vital sink for greenhouse gases, whereas land conversion to agriculture and pasture transforms the land into a source of greenhouse gases. Additionally, land-use change significantly impacts the soil's nutrient metabolism.

2016). These microbial processes significantly contribute to N availability (Camenzind et al., 2018).

However, land-use change in the Amazon region notably disrupts the N cycle (Merloti et al., 2019; Pedrinho et al., 2020). This is primarily due to the high sensitivity of soil microorganisms to environmental disturbances (Mendes et al., 2015b). Therefore, when tropical forests are converted into pasture or agricultural fields, the processes mediated by soil microorganisms undergo significant alterations, thereby affecting N dynamics and availability. Different studies performed using eDNA sequencing have consistently demonstrated that forest-to-pasture or forest-to-agriculture conversion in the Amazon region can alter the structure, diversity, and composition of the soil microbial communities, subsequently affecting their functionality (Goss-Souza et al., 2017; Mendes et al., 2015b; Merloti et al., 2019; Pedrinho et al., 2019, 2020).

For example, biological nitrogen fixation (BNF) transforms atmospheric dinitrogen gas (N_2) into ammonia (NH_3), a vital process supplying new nitrogen to terrestrial ecosystems. BNF relies on free-living or symbiotic diazotrophic microorganisms that contain the *nif*H gene, encoding the reductase subunit of nitrogenase, the enzyme responsible for this conversion (Pajares and Bohannan, 2016). Previous studies in the Amazon region have highlighted that forest-to-pasture or forest-to-agriculture conversions induce changes in the structure, diversity, composition, and functioning of these soil microorganisms (Merloti et al., 2019; Mirza et al., 2014; Pedrinho et al., 2020). Mirza et al. (2014) and Pedrinho et al. (2020) observed a significant increase in the abundance of the *nif*H gene in pasture soils post-forest conversion compared to forest soils. Similarly, Merloti et al. (2019) noted a substantial rise in the abundance of the *nif*H gene in agricultural soils following forest-to-agriculture conversion. Such changes in *nif*H gene abundance and nitrogen fixation generally reflect the response of diazotrophic microorganisms to alterations in land use, plant community composition, and management practices (Danielson and Rodrigues, 2022). These alterations may contribute to an increase (Navarrete et al., 2015), decrease (Merloti et al., 2019; Pedrinho et al., 2020), or stability (Durigan et al., 2017) of soil carbon and nitrogen levels over time.

Nitrification is a multistep reaction in which ammonium (NH_4^+) is converted to nitrate (NO_3^-). This process plays a crucial role in the N cycle of terrestrial ecosystems as it regulates nitrogen availability and loss. Nitrification is facilitated by various microorganisms, including ammonia-oxidizing archaea (AOA) and bacteria (AOB), nitrite-oxidizing bacteria (NOB), and heterotrophic bacteria and fungi capable of utilizing organic nitrogen as a substrate (Pajares and Bohannan, 2016). These ammonia-oxidizing microorganisms (AOMs) carry the *amo*A gene, encoding the α-subunit of ammonia monooxygenase, the enzyme responsible for catalyzing this reaction. According to Zhang et al. (2016), changes in soil physicochemical properties, such as NH_4^+ concentration, soil pH, temperature, moisture, etc., significantly impact AOMs (both AOA and AOB). Consequently, the conversion of tropical forests to pasture or agricultural fields can lead to dramatic alterations in ammonia-oxidizer structure, composition, and activity (Hamaoui Jr et al., 2016; Merloti et al., 2019). Previous studies conducted in the Amazon region have indeed indicated that forest-to-pasture or forest-to-agriculture conversions can modify the abundance of AOMs and increase net mineralization and nitrification rates (Merloti et al., 2019; Neill et al., 1997).

Denitrification, the final step in the nitrogen cycle, is regulated by several microbial groups through an anaerobic respiration pathway, converting NO_3^- and/or nitrite (NO_2^-) to nitrous oxide (N_2O) and/or N_2. The denitrification process comprises four reactions catalyzed by the metalloenzymes nitrate reductase (*nar*G and *nap*A marker-genes), nitrite reductase (*nir*K and *nir*S), nitric oxide reductase (*nor*B), and nitrous oxide reductase (*nos*Z). Like the microorganisms involved in BNF and nitrification, denitrifiers are sensitive to changes in soil physicochemical properties, such as soil moisture and NO_3^- concentration, induced by forest-to-pasture and/or forest-to-agriculture conversions (Merloti et al., 2019; Pedrinho et al., 2020; Pajares and Bohannan, 2016). Previous studies conducted in the Brazilian Amazon region have observed an increase in the abundance of *nir*K and *nos*Z genes, alongside denitrification rates, following forest-to-pasture and/or forest-to-agriculture conversions (Merloti et al., 2019; Pedrinho et al., 2020). This suggests that such conversions lead to a system more susceptible to nitrogen losses.

3.2 Effects of land-use change on phosphorus dynamics in the Amazon region

Amazonian soils, along with many other tropical soils, contain high levels of iron (Fe) and aluminum (Al) oxides, a result of the intense weathering processes prevalent in this region, characterized by heavy rainfall and high temperatures (Pedrinho et al., 2023; Soltangheisi et al., 2019). These Fe and Al oxides possess geochemical properties enabling them to bind to phosphorus (P), rendering them unavailable for plant uptake or other soil organisms (Gama-Rodrigues et al., 2014). Consequently, many plants and soil organisms rely on the recycling of P from soil and litterfall into forms readily usable. It is important to note that this phenomenon is particularly prominent in tropical forests, where the process of P transformation is essentially closed, exhibiting minimal losses or gains (Solomon et al., 2002).

In this context, soil microorganisms, including bacteria and fungi, play a pivotal role as they participate in three principal processes that influence P availability in soils: mineralization, solubilization, and immobilization. Microorganisms engaged in P mineralization possess the capability to produce enzymes that release P from recalcitrant organic forms (Po), thereby converting them into accessible inorganic forms (Pi) for plants and other soil organisms (Arenberg and Arai, 2019). Those involved in P solubilization can generate and discharge organic acids like citric acid, gluconic acid, and malic acid, which effectively solubilize recalcitrant P forms in soils (Alori et al., 2017). Lastly, microorganisms engaged in P immobilization can assimilate inorganic P into their biomass, competing with plants and other soil microbes for the available P (Richardson and Simpson, 2011).

Recent studies have highlighted the impact of land-use changes on the microbial community involved in P transformation and their functionality. Pedrinho et al. (2023) found that forest-to-pasture conversion results in an increase in microbial groups associated with P mineralization in pasture soil. These microorganisms typically harbor genes that encode enzymes like C–P lyases (phn) and alkaline phosphatase (phoD and phoA), known for their ability to release P from recalcitrant organic forms in soils. Consequently, such land-use changes can modify the total amount of P and its fractions in soils (Pedrinho et al., 2023; Soltangheisi et al., 2019), ultimately exerting negative effects on soil P dynamics in the Amazon region (Chavarro-Bermeo et al., 2022).

3.3 Antibiotic resistance genes in land-use-altered environments

Soil microorganisms constitute the primary source of antibiotics in natural environments (Xie et al., 2018). Typically, antibiotics produced by soil microorganisms function as signaling molecules at minimal concentrations. At higher concentrations, these antibiotics serve as a mechanism for microbial 'arming,' allowing them to competitively outmatch neighboring soil organisms for resources and space (Martínez et al., 2015). However, anthropogenic activities, including land-use changes and antibiotic usage in agriculture, progressively alter these microbial communities and the types and quantities of antibiotics they produce (Lemos et al., 2021b; Xie et al., 2018). These human-induced activities also influence the emergence and dissemination of antibiotic-resistant microbial groups (Wright, 2007). The resistance capability in these microbial groups is conferred by ARGs, often transferable within and among microbial communities through horizontal gene transfer (Xie et al., 2018). Hence, understanding how anthropogenic activities, particularly those in the Amazon region, impact the spread of ARGs is crucial. Such insights are pivotal in devising practices to mitigate this global health and food security threat (Lemos et al., 2021b).

While numerous studies have assessed the effects of land-use change on soil microbial communities and functions in the Amazon region, its influence on the soil resistome, encompassing all ARGs (Wright, 2007), remains largely unexplored. To date, only one study has investigated this aspect, demonstrating that land-use change significantly impacts the structure, diversity, and abundance of ARGs (Lemos et al., 2021b). Following forest-to-pasture conversion, a notable increase in various functional mechanisms and ARGs was observed in pasture soils. Particularly concerning was the significant rise in genes associated with resistance to tetracycline, a widely used antibiotic in livestock husbandry for treating animal diseases (He et al., 2020). This poses a substantial concern, as anthropogenic activities in the Amazon region have the potential to markedly alter the diversity of ARGs in the soil reservoir, potentially entering the food chain through the consumption of meat, vegetables, and other food sources (Cheng et al., 2019; Lemos et al., 2021b).

3.4 Effects of land-use change on greenhouse gas emissions in the Amazon region

The Amazon rainforest functions primarily as a carbon sink, sequestering carbon dioxide (CO_2) from the atmosphere through photosynthesis and storing it in both aboveground (trees and vegetation) and belowground compartments, including dead wood, litter, and soil (Mitchard, 2018; Singh et al., 2020). However, anthropogenic activities like deforestation and fires lead to the release of stored carbon, resulting in increased emissions of CO_2 and methane (CH_4) into the atmosphere (Baccini et al., 2017). These disturbances disrupt the carbon balance, altering the ratio between photosynthesis and respiration and consequently impacting soil microorganisms, such as archaea, bacteria, and fungi, which play crucial roles in carbon cycling (Gougoulias et al., 2014; Singh et al., 2020).

These soil microorganisms play a pivotal role in both producing and consuming GHGs, including carbon dioxide (CO_2), methane (CH_4), and nitrous oxide (N_2O), within terrestrial ecosystems (Stein, 2020). As mentioned earlier, these microorganisms break down organic matter into forms accessible to plants and other soil organisms (Elrys et al., 2023). In this process, they release CO_2 and methane CH_4 into the atmosphere as byproducts (Stein, 2020). However, land-use changes, especially forest-to-pasture or forest-to-agriculture conversions, can disrupt the activity, composition, and functionality of these soil microbial communities. Such disturbances negatively impact carbon decomposition rates, increase GHG emissions, and reduce the Amazonian soils' capacity to act as a carbon sink, potentially transforming them into a carbon source, thereby intensifying climate change (Danielson and Rodrigues, 2022; Obregon Alvarez et al., 2023; Steudler et al., 1996).

Several studies conducted in the Amazon region have reported divergent effects on soil carbon levels after forest-to-pasture conversion: an increase (Durrer et al., 2021), a decrease (Maia et al., 2010), or no significant change (Durigan et al., 2017). This variability suggests that the influence of forest-to-pasture or forest-to-agriculture conversions on soil carbon is contingent upon pre-existing conditions. As highlighted by Danielson and Rodrigues (2022), these conditions might be linked to factors such as initial carbon stocks, soil characteristics (e.g. texture, nutrient content), and land management practices (e.g. fires, grazing intensity, fertilization). The impact of land-use change on CH_4 flux in the Amazon has received greater attention compared to CO_2, mainly due to the involvement of a smaller number of genes and microorganisms in CH_4 cycling, and the significantly higher global warming potential of CH_4, which is 27–30 times greater than CO_2 (IPCC, 2014). The flux of CH_4 depends on how much is made versus how much is used by microorganisms. Under anaerobic conditions, methanogenic archaea produce CH_4 using the *mcr*A gene, converting CO_2 and methanol (CH_3OH) (Conrad, 2007). In contrast, under aerobic conditions, methanotrophic bacteria, possessing the *pmo*A gene, turn CH_4 into methanol using O_2 as an electron acceptor (Conrad, 2007).

Studies conducted in the Amazon region highlight the significant impact of forest-to-pasture conversion on both methanogenic archaea and methanotrophic bacteria. After the transition from forest to pasture, there is a notable increase in the abundance and diversity of methanogenic archaea in pasture soils, as noted by Obregon Alvarez et al. (2023). Additionally, observations by Meyer et al. (2017) and Paula et al. (2014) indicate that forest-to-pasture conversion leads to a decline in the abundance and alters the composition of methanotrophic bacteria in pasture soils. Consequently, this change in land use consistently affects CH_4 flux across the Amazon basin. Research suggests that tropical forests usually function as a sink for CH_4 (−470 mg CH_4 m²), while pastures serve as a source (+270 mg CH_4 m²) (Danielson and Rodrigues, 2022; Steudler et al., 1996).

It is also crucial to highlight N_2O as another significant GHG, with a global warming potential 300 times higher than CO_2 (Forster et al., 2007; UNEP, 2013). As discussed previously, N_2O is generated during the denitrification process by a group of microorganisms known as denitrifiers. A meta-analysis of N_2O flux studies across the Amazon basin reveals that tropical forests generally exhibit higher annual emissions (2.42 kg N ha⁻¹) compared to old pasture soils (0.9 kg N ha⁻¹), which, in certain cases, may even function as a slight sink (Meurer et al., 2016). However, the same analysis by Meurer et al. (2016) noted that recently converted pasture areas (less than 10 years) present elevated annual emissions (2.52 kg N ha⁻¹), akin to levels observed in tropical forests. Moreover, it is worth noting that appropriate pasture management can also stimulate methane capture by the soil. A recent study conducted in the Amazon has shown that better management of grasses in pastures can enhance the soil's capacity for carbon sequestration, mainly due to an increase in the abundance of methanotrophic microorganisms in the rhizosphere of the plants (Souza et al., 2022).

4 Effects of land recovery on soil microbiome properties

Extensive land use for agriculture and pasture often leads to soil degradation, a process detrimental to the health and fertility of the land. Continuous cultivation of crops or grazing in pastures without adequate rest or sustainable management practices can strip the soil of essential nutrients, decrease organic matter, and erode its structure. This overexploitation diminishes soil quality, reducing its ability to retain water, support plant growth, and maintain biodiversity. Land degradation poses a significant obstacle to sustainable development, with far-reaching consequences for food security, biodiversity conservation, and climate change mitigation and adaptation. Therefore, urgent action is required to devise strategies aimed at reducing and restoring degraded ecosystems. There are several strategies to reduce and restore degraded land and ecosystems, including conservation, reforestation, and regenerative agriculture. Figure 3 illustrates the results discussed in this section.

4.1 Land recovery through secondary forest

Typically, the conversion of forest to pasture in the Amazon region follows a process involving selective logging of valuable timber, subsequent burning of

Figure 3 Strategies for enhancing ecosystem recovery. The figure illustrates diverse approaches utilized to mitigate land degradation and promote restoration, with particular emphasis on their effects on the microbiome and the environment.

the remaining vegetation, and the seeding of forage grasses (*Urochloa* genera) (Meyer et al., 2017; Navarrete et al., 2015; Pedrinho et al., 2019). However, within 5–15 years of exploitation and poor management practices, pastures become degraded and unproductive and, consequently, are abandoned (Souza Braz et al., 2013; Pedrinho et al., 2019; Pedrinho et al., 2020). As a result, the secondary forest vegetation starts to build up a new forest stand (Cenciani et al., 2009; Pedrinho et al., 2020). Several studies performed in the Amazon region have compared microbial communities in primary forests and pastures with those in secondary forests (Durrer et al., 2021; Paula et al., 2014; Pedrinho et al., 2019; Pedrinho et al., 2020; Pedrinho et al., 2023). Many of these studies concluded that microbial diversity indices in secondary forests are more similar to primary forests and that specific microbial groups and functions can be recovered over time (especially those linked to nutrient cycling) (Paula et al., 2014; Pedrinho et al., 2020; Pedrinho et al., 2023). For example, when focusing on the nitrogen cycle, different studies suggested that, during secondary forest succession, the abundance of diazotrophs decreases in comparison to pastures, converging toward levels observed in primary forests (Mirza et al., 2014; Pedrinho et al., 2020). The same pattern is observed for the abundance of the *amoA* gene (maker gene of AOMs) in secondary forest soil, which presented values similar to those observed in primary forests (Hamaoui et al., 2016).

For instance, numerous studies have shown that the conversion of natural forest areas into pasturelands disrupts the methane gas flow in Amazonian soils, further contributing to the problem of climate change (Meyer et al., 2017; Kroeger et al., 2021; Venturini et al., 2022; Alvarez et al., 2023). However, the recovery of degraded areas through the growth of secondary forests has shown that microbial functions have started to return to the state of native vegetation (Pedrinho et al., 2019, 2020, 2023). Moreover, it is worth noting that appropriate pasture management can also stimulate methane capture by the soil. A recent study conducted in the Amazon has shown that better management of grasses in pastures can enhance the soil's capacity for carbon sequestration, mainly due to an increase in the abundance of methanotrophic microorganisms in the rhizosphere of the plants (Souza et al., 2022). These examples illustrate how the recovery of vegetation and the proper management of agricultural areas can effectively balance GHG emissions by altering the abundance of specific microorganisms involved in the carbon cycle. Considering that microbial diversity is essential in determining the global biogeochemical balance, improved management practices along with microbiome manipulation techniques can serve as a powerful tool to accelerate the recovery of degraded areas and reduce the impact of climate change.

4.2 Land recovery through conservationist practices in agriculture

Conservationist practices in agriculture is a sustainable farming system that aims to conserve, improve, and efficiently utilize natural resources through integrated management of soil, water, and biological resources combined with external inputs. It involves minimizing soil disturbance, improving organic matter and soil cover, and using crop rotations and associations to reduce the impact of pests and diseases. Practices such as no-tillage systems, cover crops, terracing, and agroforestry aim to maintain soil, water, and biodiversity, and reduce soil erosion and compaction. These practices also positively impact the soil microbiome, influencing the composition and functioning of bacterial and fungal communities, and improving soil health (Banerjee et al., 2018; Tao et al., 2020). Furthermore, conservationist practices can promote disease resistance-inducing and growth-promoting beneficial microbes in the soil, potentially enhancing plant survival and crop production (Berendsen et al., 2018; Raaijmakers & Mazzola, 2016; Tao et al., 2020). However, the precise functional role of the soil microbiome in agricultural systems is still not fully understood, highlighting the need for further research to optimize conservationist practices for improved crops (Zhao et al., 2020). Overall, the integration of conservationist practices in agriculture is crucial for sustainable soil management and crop production, emphasizing the importance of understanding and managing the soil microbiome in agricultural systems (Zhao et al., 2020; Yuan et al., 2021).

No-till practices in agriculture have been shown to have a significant impact on soil microbiomes. These practices help maintain nutrients within the soil, reducing the need for supplements (Smith et al., 2016). Furthermore, no-till methods are critical for restoring and protecting soil health, contributing to the sustainability of global agriculture (Anderson, 2015). Additionally, agricultural practices such as tillage and cropping systems can lead to structural and functional changes in soil microbiomes, affecting soil health and plant disease resistance (Foo et al., 2017). The initial soil microbiome composition and functioning can predetermine future plant health, highlighting the long-term impact of agricultural practices on soil microbiomes (Wei et al., 2019). Overall, the adoption of no-till practices not only provides economic and environmental benefits but also contributes to the development of disease-suppressive soil microbiomes, supporting sustainable agriculture (Chen et al., 2022).

Another example of conservationist practices is crop rotation and cover cropping. These are agricultural practices that have demonstrated the ability to induce changes in soil microbiomes. These practices increase soil fertility, microbial diversity, and disease-suppressive capacity, leading to improved soil health and crop productivity (Castellano-Hinojosa & Strauss, 2020; Yuan et al., 2021). The intentional management of soil microbiomes through cover crop selection can be a strategy to enhance soil health and promote beneficial changes in soil chemical and biological attributes, such as increasing nitrogen and carbon availability and microbial diversity. Furthermore, crop rotation has been found to be effective in maintaining microbiome complexity and functioning in salinity-affected soils and in reducing the negative effects of salinity on soil microbiomes (Dasgupta, 2023). These practices have also been shown to regulate nutrition, reduce soilborne diseases, and improve soil fertility, making them important strategies for sustainable agriculture (Yuan et al., 2021).

Agroforestry practices, another conservation method, have been shown to alter soil microbial diversity and composition. This primarily occurs through the creation of a distinct tree row-associated microbiome, which differs from the crop row microbiome, thereby increasing overall beta diversity (Fahad et al., 2022). Additionally, agroforestry systems can promote diversity, alter the composition, and improve the resilience of root-associated fungal communities, which can have a positive feedback effect on the surrounding soil after the long-term establishment of tree enrichment (Ballauff et al., 2020). Furthermore, management practices in agro-systems generate spatial and temporal changes in soil physical and chemical properties, creating rapidly fluctuating environments that provide a wide range of niches for microbial growth (Dube et al., 2019). The impact of agroforestry on soil microbiomes is also influenced by factors such as nutrient input, which appears to be a major factor fostering the differentiation of soil microbiome between tree rows and arable land in agroforestry systems (Beule & Karlovsky, 2021). Moreover, the manipulation of the microbiome through agroforestry practices may reduce the need for fertilizer and pesticide use, leading to more sustainable agricultural practices (Liu et al., 2023).

Notably, conservationist practices in agriculture have immense potential for enhancing soil health, promoting biodiversity, and increasing crop productivity. These practices not only minimize soil erosion and compaction but also exert a significant influence on the soil microbiome. The microbiome plays a crucial role in nutrient cycling, disease suppression, and overall soil fertility, highlighting its significance in sustainable agriculture. In the pursuit of sustainable agriculture, integrating conservationist practices and managing soil microbiomes carefully are essential components. By harnessing the power of beneficial microbial communities and fostering harmonious relationships between crops and their microbial allies, we can pave the way toward a more resilient and environmentally friendly agricultural future.

5 Conclusion

Land-use management has a significant impact on soil microbiomes, influencing their composition, diversity, and functions. The changes in land-use types and agriculture practices significantly affect the variability of microbiomes in soil profiles. These changes can lead to distinct community assemblies and functions of microbiomes throughout soil profiles, highlighting the need for strategies and policies that consider the responses of vertical soil microbiomes to land-use changes. Furthermore, the role of land use in structuring the microbiomes of connected ecosystems has been emphasized, indicating that land-use types and water chemical properties play a crucial role in shaping the microbiomes of interconnected environments (Marmen et al., 2020).

Advanced molecular techniques, such as Next-Generation Sequencing, have provided a deeper understanding of the soil microbiome and its responses to land-use management. However, it is important to note that inherent biases in soil metagenomics present challenges in accurately defining the soil microbiome and its ecosystem function (Lemos et al., 2021a; Sanchez-Cid et al., 2022). Additionally, specialized metabolic functions of keystone taxa have been found to sustain soil microbiome stability, indicating the potential of specific microbial groups to maintain stability in the face of environmental perturbations caused by land-use management and other factors (Xun et al., 2021). Microbiome-based predictive models have been developed to understand how soil microbial biodiversity might be affected by future climate change scenarios, providing valuable insights into the potential responses of soil microbiomes to changing environmental conditions (Cowan et al., 2022). Moreover, the functional stability of the soil microbiome has been linked to the maintenance of ecosystem services such as primary production, soil carbon accumulation, and nutrient cycling, highlighting the importance of understanding and predicting the responses of soil microbiomes to land use management for ecosystem sustainability (Jiao et al., 2021).

Considering that climate change is pivotal in assessing the future perspectives of how land-use management affects soil microbiomes, it is essential to recognize that long-term shifts in average weather patterns can profoundly impact the soil microbiome and biodiversity. This emphasizes the necessity of acknowledging climate change as a significant factor influencing how soil microbiomes respond to land-use management practices (Mishra et al., 2022). Additionally, the responses of soil microbiomes to environmental changes display notable geographic patterns, indicating the intricate interplay among land use, climate, and the dynamics of soil microbiomes (Jiao et al., 2022).

In conclusion, future perspectives on how land-use management affects soil microbiomes are multifaceted, encompassing advanced molecular techniques, microbiome-based predictive models, climate change considerations, and ecosystem services assessment. Understanding the intricate relationships between land-use management and soil microbiomes is essential for sustainable land management practices and the preservation of ecosystem functions.

6 Where to look for further information

- Danielson RE, Rodrigues JL (2022) Impacts of land-use change on soil microbial communities and their function in the Amazon Rainforest. *Advances in Agronomy* 175:179–258. https://doi.org/10.1016/bs.agron .2022.04.001.
- Horwath W (Ed.) (2022) *Improving Soil Health*. Sawston: Burleigh Dodds Science Publishing.
- Lambin EF, Geist H (Eds.) (2010) *Land-Use and Land-Cover Change: Local Processes and Global Impacts*. Berlin and Heidelberg: Springer. https:// doi.org/10.1007/3-540-32202-7.
- Otten W (Ed.) *Advances in Measuring Soil Health*. Burleigh Dodds Science Publishing. https://doi.org/10.1201/9781003048046.
- Pedrinho A, Mendes LW, Pereira APA, Araujo ASF, Vaishnav A, Karpouzas DG, Singh BK (2024) Soil microbial diversity plays an important role in resisting and restoring degraded ecosystems. *Plant and Soil*. https://doi .org/10.1007/s11104-024-06489-x.
- Pylro V, Roesch L (Eds.) (2017) *The Brazilian Microbiome: Current Status and Perspectives*. Cham: Springer. https://doi.org/10.1007/978-3-319-59997-7.
- Reicosky D (Ed.) (2018a) *Managing Soil Health for Sustainable Agriculture. Volume 1: Fundamentals*. Burleigh Dodds Science Publishing. https://doi .org/10.1201/9781351114530.
- Reicosky D (Ed.) (2018b) *Managing Soil Health for Sustainable Agriculture. Volume 2: Monitoring and Management*. Burleigh Dodds Science Publishing. https://doi.org/10.1201/9781351114585.
- Singh JS, Tiwari S, Singh C, Singh AK (Eds.) (2021) *Microbes in Land Use Change Management*. Elsevier. https://doi.org/10.1016/C2020-0-00650 -X.
- van Elsas JD, Trevors JT, Soares Rosado A, Nannipieri P (Eds.) (2019) *Modern Soil Microbiology, Third Edition* (3rd ed.). CRC Press. https://doi .org/10.1201/9780429059186.
- Yadav AN (Ed.) (2021) *Soil Microbiomes for Sustainable Agriculture: Functional Annotation*. Cham: Springer. https://doi.org/10.1007/978-3 -030-73507-4.

7 References

Alori ET, et al. (2017) Microbial phosphorus solubilization and its potential for use in sustainable agriculture. *Frontiers in Microbiology* 8, 971.

Anderson, R (2015) Integrating a complex rotation with no-till improves weed management in organic farming. A review. *Agronomy for Sustainable Development* 35(3), 967–974. https://doi.org/10.1007/s13593-015-0292-3

Arenberg MR, Arai Y (2019) Uncertainties in soil physicochemical factors controlling phosphorus mineralization and immobilization processes. *Advances in Agronomy* 154, 153–200. https://doi.org/10.1016/bs.agron.2018.11.005

Averill C, et al. (2022) Defending Earth's terrestrial microbiome. *Nature Microbiology* 7, 1717–1725.

Baccini A, et al. (2017) Tropical forests are a net carbon source based on aboveground measurements of gain and loss. *Science* 358(6360), 230–234. https://doi.org/10.1126/science.aam5962

Bach EM, et al. (2020) Soil biodiversity integrates solutions for a sustainable future. *Sustainability* 12, 2662.

Ballauff J, et al. (2020) Legacy effects overshadow tree diversity effects on soil fungal communities in oil palm-enrichment plantations. *Microorganisms* 8(10), 1577. https://doi.org/10.3390/microorganisms8101577

Banerjee S, et al. (2018) Keystone taxa as drivers of microbiome structure and functioning. *Nature Reviews Microbiology* 16(9), 567–576. https://doi.org/10.1038/s41579-018-0024-1

Berendsen R, et al. (2018) Disease-induced assemblage of a plant-beneficial bacterial consortium. *The Isme Journal*, 12(6), 1496–1507. https://doi.org/10.1038/s41396-018-0093-1

Beule L, Karlovsky, P (2021) Tree rows in temperate agroforestry croplands alter the composition of soil bacterial communities. *PLoS One*, 16(2), e0246919. https://doi.org/10.1371/journal.pone.0246919

Borneman J, Triplett EW (1997) Molecular microbial diversity in soils from Eastern Amazonia: evidence for unusual microorganisms and microbial population shifts associated with deforestation. *Applied and Environmental Microbiology* 63, 2647–2653.

Brossi MJL, et al. (2014) Assessment of bacterial bph gene in Amazon Dark Earth and their adjacent soils. *PLoS One* 9, e99597.

Camenzind T, et al. (2018) Nutrient limitation of soil microbial processes in tropical forests. *Ecological Monographs* 88(1), 4–21. https://doi.org/10.1002/ecm.1279

Cardinale BJ, et al (2012) Biodiversity loss and its impact on humanity. Nature, 486(7401), 59–67. doi: 10.1038/nature11148

Castellano-Hinojosa, A. and Strauss, S (2020) Impact of cover crops on the soil microbiome of tree crops. *Microorganisms* 8(3), 328. https://doi.org/10.3390/microorganisms8030328

Cenciani K, et al. (2009) Bacteria diversity and microbial biomass in forest, pasture and fallow soils in the southwestern Amazon basin. *Revista Brasileira de Ciência do Solo* 33, 907–916. https://doi.org/10.1590/S0100-06832009000400015

Ceola G, et al. (2021) Biogeographic patterns of arbuscular mycorrhizal fungal communities along a land-use intensification gradient in the subtropical atlantic forest biome. *Microbial Ecology*. https://doi.org/10.1007/s00248-021-01721-y

Chavarro-Bermeo JP, et al. (2022) Responses of soil phosphorus fractions to land-use change in Colombian Amazon. *Sustainability* 14(4), 2285. https://doi.org/10.3390/su14042285

Chen L, et al. (2022) The impact of no-till on agricultural land values in the united states midwest. *American Journal of Agricultural Economics* 105(3), 760–783. https://doi .org/10.1111/ajae.12338

Cheng G, et al. (2019) Selection and dissemination of antimicrobial resistance in Agri-food production. *Antimicrobial Resistance and Infection Control* 8, 158. https://doi .org/10.1186/s13756-019-0623-2

Conrad R (2007) Microbial ecology of methanogens and methanotrophs. *Advances in Agronomy* 96, 1–63. https://doi.org/10.1016/S0065-2113(07)96005-8

Cowan D, et al. (2022) Biogeographical survey of soil microbiomes across sub-saharan africa: structure, drivers, and predicted climate-driven changes. *Microbiome* 10(1).https://doi.org/10.1186/s40168-022-01297-w

Danielson RE, Rodrigues JL (2022) Impacts of land-use change on soil microbial communities and their function in the Amazon Rainforest. *Advances in Agronomy* 175, 179–258. https://doi.org/10.1016/bs.agron.2022.04.001

Dasgupta D (2023) Cover cropping reduces the negative effect of salinity on soil microbiomes. *Journal of Sustainable Agriculture and Environment* 2(2), 140–152. https://doi.org/10.1002/sae2.12054

Delory BM, et al. 2019. When history matters: the overlooked role of priority effects in grassland overyielding. *Functional Ecology* 33, 2369–2380. https://doi.org/10.1111 /1365-2435.13455.

Dube J, et al. (2019) Differences in bacterial diversity, composition and function due to long-term agriculture in soils in the eastern free state of south africa. *Diversity* 11(4), 61. https://doi.org/10.3390/d11040061

Durigan M, et al. (2017) Soil organic matter responses to anthropogenic forest disturbance and land use change in the Eastern Brazilian Amazon. *Sustainability* 9, 379. https:// doi.org/10.3390/su9030379

Durrer A, et al. (2021) Beyond total carbon: conversion of amazon forest to pasture alters indicators of soil C cycling. *Biogeochemistry* 152, 179–194. https://doi.org/10.1007 /s10533-020-00743

Elrys AS, et al. (2023) Global soil nitrogen cycle pattern and nitrogen enrichment effects: tropical versus subtropical forests. *Global Change Biology* 29(7), 1905-1921. https:// doi.org/10.1111/gcb.16603

Fahad S, et al. (2022) Agroforestry systems for soil health improvement and maintenance. *Sustainability*, 14(22), 14877. https://doi.org/10.3390/su142214877

Faoro H, et al. (2010) Influence of soil characteristics on the diversity of bacteria in the southern brazilian atlantic forest. *Applied and Environmental Microbiology* 76, 4744–4749. https://doi.org/10.1128/AEM.03025-09

Foo J, et al. (2017) Microbiome engineering: current applications and its future. *Biotechnology Journal* 12(3).https://doi.org/10.1002/biot.201600099

Forster P, et al. (2007) Changes in atmospheric constituents and in radiative forcing. In: Solomon, S., et al. (ed.) *Climate Change: The Physical Science Basis. International Nuclear Information System.* Cambridge University Press, Cambridge, United Kingdom and New York, NY, USA, 129–234.

Fujii K, et al. (2018) Plant–soil interactions maintain biodiversity and functions of tropical forest ecosystems. *Ecological Research* 33, 149–160. https://doi.org/10.1007/ s11284-017-1511-y

Gama-Rodrigues AC, et al. (2014) An exploratory analysis of phosphorus transformations in tropical soils using structural equation modeling. *Biogeochemistry* 118, 453–469. https://doi.org/10.1007/s10533-013-9946-x

Germano MG, et al. (2012) Functional diversity of bacterial genes associated with aromatic hydrocarbon degradation in anthropogenic dark earth of Amazonia. *Pesquisa Agropecuaria Brasileira* 47, 654–664.

Goss-Souza D, et al. (2017) Soil microbial community dynamics and assembly under long-term land use change. *FEMS Microbiology Ecology* 93(10), fix109. https://doi.org/10.1093/femsec/fix109

Goss-Souza D, et al. (2019) Amazon forest-to-agriculture conversion alters rhizosphere microbiome composition while functions are kept. *FEMS Microbiology Ecology.* https://doi.org/10.1093/femsec/fiz009

Goss-Souza D, et al. (2020) Ecological processes shaping bulk soil and rhizosphere microbiome assembly in a long-term amazon forest-to- agriculture conversion. *Microbial Ecology* 79, 110–122. https:// doi.org/10.1007/s00248-019-01401-y

Gougoulias C, et al. (2014) The role of soil microbes in the global carbon cycle: tracking the below-ground microbial processing of plant-derived carbon for manipulating carbon dynamics in agricultural systems. *Journal of the Science of Food and Agriculture* 94(12), 2362–2371. https://doi.org/10.1002/jsfa.6577

Grossman JM, et al. (2010) An assessment of nodulation and nitrogen fixation in inoculated Inga oestediana, a nitrogen-fixing tree shading organically grown coffee in Chiapas, Mexico. *Soil Biology and Biochemistry* 38, 769–784.

Hamaoui Jr GS, et al. (2016) Land-use change drives abundance and community structure alterations of thaumarchaeal ammonia oxidizers in tropical rainforest soils in Rondônia, Brazil. *Applied Soil Ecology* 107, 48–56. https://doi.org/10.1016/j.apsoil.2016.05.012

He Y, et al. (2020) Antibiotic resistance genes from livestock waste: occurrence, dissemination, and treatment. *NPJ Clean Water* 3, 1–11. https://doi.org/10.1038/s41545-020-0051-0

Intergovernmental Panel on Climate Change – IPCC (2014) Climate change 2014: synthesis report. Pachauri RK, Allen MR, Barros VR, Broome J, Cramer W, Christ R, Church JA, Clarke L, Dahe Q, Dasgupta P, Dubash NK (eds) Contribution of Working Groups I, II and III to the fifth assessment report of the Intergovernmental Panel on Climate Change. Geneva, p. 151.

Jantz SM, et al. (2015) Future habitat loss and extinctions driven by land-use change in biodiversity hotspots under four scenarios of climate-change mitigation. *Conservation Biology* 29, 1122–1131. https://doi.org/10.1111/cobi.12549

Jesus EC, et al. (2009) Changes in land use alter the structure of bacterial communities in Western Amazon soils. *The ISME Journal* 3, 1004–1011. https://doi.org/10.1038/ismej.2009.47

Jiao S, et al. (2021) Core microbiota drive functional stability of soil microbiome in reforestation ecosystems. *Global Change Biology*, 28(3), 1038–1047. https://doi.org/10.1111/gcb.16024

Jiao S, et al. (2022) Core phylotypes enhance the resistance of soil microbiome to environmental changes to maintain multifunctionality in agricultural ecosystems. *Global Change Biology*, 28(22), 6653–6664. https://doi.org/10.1111/gcb.16387

Kaiser K, et al. (2016) Driving forces of soil bacterial community structure, diversity, and function in temperate grasslands and forests. *Scientific Reports* 6, 1–12. https://doi.org/10.1038/srep33696

Kroeger M. E., et al. (2021). Rainforest-to-pasture conversion stimulates soil methanogenesis across the Brazilian Amazon. *ISME Journal* 15:658–672. https://doi .org/10.1038/s41396-020-00804-x

Lal R (2020) Soil degradation as a reason for inadequate human nutrition: significance, causes, and remedies. *Sustainability* 12(6), 2336. doi: 10.3390/su12062336

Lauber CL, et al. (2013) Temporal variability in soil microbial communities across land-use types. *The ISME Journal* 7, 1641–1650. https://doi.org/10.1038/ ismej.2013.50

Lemos LN, et al. (2021a) Genome-resolved metagenomics is essential for unlocking the microbial black box of the soil. *Trends in Microbiology* 29, 279–282.

Lemos LN, et al. (2021b) Amazon deforestation enriches antibiotic resistance genes. *Soil Biology and Biochemistry* 153, 108110. https://doi.org/10.1016/j.soilbio.2020.108110

Liu J, et al. (2023) Integrated microbiome and metabolomics analysis reveal a closer relationship between soil metabolites and bacterial community than fungal community in pecan plantations. *Land Degradation and Development* 34(10), 2812–2824. https://doi.org/10.1002/ldr.4649

Lupatini M, et al. 2019. Moisture is more important than temperature for assembly of both potentially active and whole prokaryotic communities in subtropical grassland. *Microbial Ecology* 77, 460–470. https://doi.org/10.1007/s00248-018-1310-1.

Maia SM, et al. (2010) Soil organic carbon stock change due to land use activity along the agricultural frontier of the southwestern Amazon, Brazil, between 1970 and 2002. *Global Change Biology* 16(10), 2775–2788. https://doi.org/10.1111/j.1365-2486 .2009.02105.x

Marmen S, et al. (2020) The role of land use types and water chemical properties in structuring the microbiomes of a connected lake system. *Frontiers in Microbiology* 11. https://doi.org/10.3389/fmicb.2020.00089

Martínez JL, et al. (2015) What is a resistance gene? Ranking risk in resistomes. *Nature Reviews Microbiology* 13(2), 116–123. https://doi.org/10.1038/nrmicro3399

Melo VF, et al. (2017) Land use and changes in soil morphology and physical-chemical properties in Southern Amazon. *Revista Brasileira de Ciência do Solo* 41, e0170034. https://doi.org/10.1590/18069657rbcs20170034

Mendes LW, et al. (2017) Using metagenomics to connect microbial Community biodiversity and functions. *Current Issues in Molecular Biology* 24, 103–118.

Mendes LW, et al. (2014) Taxonomical and functional microbial community selection in soybean rhizosphere. *The ISME Journal* 8, 1577–1587.

Mendes LW, et al. (2015a) Land-use system shapes soil bacterial communities in Southeastern Amazon region. *Applied Soil Ecology* 95, 151–160. https://doi.org/10 .1016/j.apsoil.2015.06.005

Mendes LW, et al. (2015b) Soil-borne microbiome: linking diversity to function. *Microbial Ecology* 70, 255–265. https://doi.org/10.1007/s00248-014-0559-2

Merloti LF, et al. (2019) Forest-to-agriculture conversion in Amazon drives soil microbial communities and N-cycle. *Soil Biology and Biochemistry* 137:107567. https://doi .org/10.1016/j.soilbio.2019.107567

Meurer KH, et al. (2016) Direct nitrous oxide (N_2O) fluxes from soils under different land use in Brazil - a critical review. *Environmental Research Letters* 11(2), 023001. https:// doi.org/10.1088/1748-9326/11/2/023001

Meyer KM, et al. (2017) Conversion of Amazon rainforest to agriculture alters community traits of methane-cycling organisms. *Molecular Ecology* 26, 1547–1556. https://doi .org/10.1111/mec.14011

Mishra A, et al. (2022) Unboxing the black box—one step forward to understand the soil microbiome: a systematic review. *Microbial Ecology* 85(2), 669–683. https://doi.org /10.1007/s00248-022-01962-5

Mirza B. S., et al. (2014). Response of Free-Living Nitrogen-Fixing Microorganisms to Land Use Change in the Amazon Rainforest. *Appl Environ Microbiol* 80. https://doi.org/10 .1128/AEM.02362-13

Mitchard ETA (2018) The tropical forest carbon cycle and climate change. *Nature* 559, 527–534. https://doi.org/10.1038/s41586-018-0300-2

Myers N, et al. (2000) Bio-diversity hotspots for conservation priorities. *Nature* 403, 853– 558. https://doi.org/10.1038/35002501.

Nascimento N, et al. (2019) What drives intensification of land use at agricultural frontiers in the Brazilian Amazon? Evidence from a decision game. *Forests* 10(6), 464. https:// doi.org/10.3390/f10060464

Navarrete AA, et al. (2011) Land-use systems affect archaeal community structure and functional diversity in Western Amazon soils. *Revista Brasileira de Ciência do Solo* 35, 1527–1540.

Navarrete AA, et al. (2015) Soil microbiome responses to the short-term effects of Amazonian deforestation. *Molecular Ecology* 24(10), 2433–2448. https://doi.org/10 .1111/mec.13172

Neill C, et al. (1997) Soil carbon and nitrogen stocks following forest clearing for pasture in the southwestern Brazilian Amazon. *Ecological Applications* 7(4), 1216-1225. https://doi.org/10.1890/1051-0761(1997)007[1216:SCANSF]2.0.CO;2

Nunes CA, et al. (2022) Linking land-use and land-cover transitions to their ecological impact in the Amazon. *Proceedings of the National Academy of Sciences* 119(27), e2202310119. https://doi.org/10.1073/pnas.2202310119

Obregon Alvarez D, et al. (2023) Shifts in functional traits and interactions patterns of soil methane-cycling communities following forest-to-pasture conversion in the Amazon Basin. *Molecular Ecology* 32(12), 3257–3275. https://doi.org/10.1111/mec .16912

Pajares S, Bohannan BJ (2016) Ecology of nitrogen fixing, nitrifying, and denitrifying microorganisms in tropical forest soils. *Frontiers in Microbiology* 7, 1045. https://doi .org/10.3389/fmicb.2016.01045

Paula FS, et al. (2014) Land use change alters functional gene diversity, composition and abundance in Amazon forest soil microbial communities. *Molecular Ecology* 23(12), 2988–2999. https://doi.org/10.1111/mec.12786

Pazinato JM, et al. (2010) Molecular characterization of the archaeal community in an Amazonian Wetland Soil and culture-dependent isolation of methanogenic Archaea. *Diversity* 2, 1026–1047.

Pedrinho A, et al. (2019) Forest-to-pasture conversion and recovery based on assessment of microbial communities in Eastern Amazon rainforest. *FEMS Microbiology Ecology* 95(3), fiy236. https://doi.org/10.1093/femsec/fiy236

Pedrinho A, et al. (2020) The natural recovery of soil microbial community and nitrogen functions after pasture abandonment in the Amazon region. *FEMS Microbiology Ecology* 96(9), fiaa149. https://doi.org/10.1093/femsec/fiaa149

Pedrinho A, et al. (2023) Impacts of deforestation and forest regeneration on soil bacterial communities associated with phosphorus transformation processes in the Brazilian Amazon region. *Ecological Indicators* 146, 109779. https://doi.org/10.1016/j .ecolind.2022.109779

Piccolo MC, et al. (1996) ^{15}N natural abundance in forest and pasture soils of the Brazilian Amazon Basin. *Plant Soil* 182, 249–258. https://doi.org/10.1007/BF00029056

Raaijmakers J, Mazzola M (2016) Soil immune responses. *Science* 352(6292), 1392–1393. https://doi.org/10.1126/science.aaf3252

Ribeiro MC, et al. (2009) The Brazilian Atlantic Forest: how much is left, and how is the remaining forest distributed? Implications for conservation. *Biological Conservation* 142, 1141–1153. https://doi.org/10. 1016/j.biocon.2009.02.021

Richardson AE, Simpson RJ (2011) Soil microorganisms mediating phosphorus availability update on microbial phosphorus. *Plant Physiology* 156(3), 989–996. https://doi.org /10.1104/pp.111.175448

Rodrigues JL, et al. (2013) Conversion of the Amazon rainforest to agriculture results in biotic homogenization of soil bacterial communities. *Proceedings of the National Academy of Sciences* 110(3), 988–993. https://doi.org/10.1073/pnas.1220608110

Sanchez-Cid C, et al. (2022) Sequencing depth has a stronger effect than dna extraction on soil bacterial richness discovery. *Biomolecules* 12(3), 364. https://doi.org/10 .3390/biom12030364

Singh A, et al. (2020) Role of microorganisms in regulating carbon cycle in tropical and subtropical soils. In: Ghosh P, et al. (eds) *Carbon Management in Tropical and Sub-Tropical Terrestrial Systems*. Singapore: Springer. https://doi.org/10.1007 /978–981-13-9628-1_15

Smith C, et al. (2016) Microbial community responses to soil tillage and crop rotation in a corn/soybean agroecosystem. *Ecology and Evolution* 6(22), 8075–8084. https://doi .org/10.1002/ece3.2553

Solomon D, et al. (2002) Phosphorus forms and dynamics as influenced by land use changes in the sub-humid Ethiopian highlands. *Geoderma* 105(1–2), 21–48. https:// doi.org/10.1016/S0016-7061(01)00090-8

Soltangheisi A, et al. (2019) Forest conversion to pasture affects soil phosphorus dynamics and nutritional status in Brazilian Amazon. *Soil and Tillage Research* 194, 104330. https://doi.org/10.1016/j.still.2019.104330

Souza LF, et al. (2022) Maintaining grass coverage increases methane uptake in Amazonian pastures, with a reduction of methanogenic archaea in the rhizosphere. *Science of the Total Environment* 838, 156225.

Souza Braz AD, et al. (2013) Soil attributes after the conversion from forest to pasture in Amazon. *Land Degradation & Development* 24(1), 33-38. https://doi.org/10.1002/ ldr.1100

Stein LY (2020) The long-term relationship between microbial metabolism and greenhouse gases. *Trends in Microbiology* 28(6), 500–511. https://doi.org/10.1016 /j.tim.2020.01.006

Steudler PA, et al. (1996) Consequence of forest-to-pasture conversion on CH_4 fluxes in the Brazilian Amazon Basin. *Journal of Geophysical Research* 101(D13), 18547–18554. https://doi.org/10.1029/96JD01551

Sühs, RB, et al. 2020. Preventing traditional management can cause grassland loss within 30 years in southern Brazil. *Scientific Reports* 10, 1–9. https://doi.org/ 10.1038/ s41598-020-57564-z

Taketani RG, et al. (2013) Bacterial community composition of anthropogenic biochar and Amazonian anthrosols assessed by 16S rRNA gene 454 pyrosequencing. *Antonie van Leeuwenhoek* 104, 233–242.

Tao C, et al. (2020) Bio-organic fertilizers stimulate indigenous soil pseudomonas populations to enhance plant disease suppression. *Microbiome* 8(1).https://doi.org /10.1186/s40168-020-00892-z

Tomazelli D, et al. (2023a) Pasture management intensification shifts the soil microbiome composition and ecosystem functions. *Agriculture, Ecosystems & Environment* 346, 108355. https://doi.org/10.1016/j.agee.2023.108355.

Tomazelli D, et al. (2023b) Goss-Souza, D. Natural grassland conversion to cultivated pastures increases soil microbial niche specialization with consequences for ecological processes. *Applied Soil Ecology* 188, 104913. https://doi.org/10.1016/j .apsoil.2023.104913.

United Nations (2017) Global Land Outlook. United Nations Convention to Combat Desertification, Bonn, Germany. https://www.unccd.int/resources/global-land -outlook

United Nations Environment Programme (UNEP) (2013) *Drawing Down N$_2$O to Protect Climate and Ozone Layer*. Nairobi, Kenya: UNEP.

Vieira ICG (2019) Land use drives change in Amazonian tree species. *Anais da Academia Brasileira de Ciências* 91, e20190186. https://doi.org/10.1590/0001 -3765201920190186

Venturini A. M., et al. (2022) Increased soil moisture intensifies the impacts of forest-to-pasture conversion on methane emissions and methane-cycling communities in the Eastern Amazon. Environmental Research 212 (Part A): 113139. https://doi.org/10 .1016/j.envres.2022.113139

Wang P, et al. (2021) Disturbances consistently restrain the role of random migration in grassland soil microbial community assembly. *Global Ecology and Conservation* 26, e01452 https://doi.org/10.1016/j. Gecco.2021.e01452.

Wei Z, et al. (2019) Initial soil microbiome composition and functioning predetermine future plant health. *Science Advances* 5(9) https://doi.org/10.1126/sciadv.aaw0759

Weinzettel J, et al. (2018) Human footprint in biodiversity hotspots. *Frontiers in Ecology and the Environment* 16, 447–452. https://doi.org/10.1002/fee.1825

Wright GD (2007) The antibiotic resistome: the nexus of chemical and genetic diversity. *Nature Reviews Microbiology* 5, 175–186. https://doi.org/10.1038/nrmicro1614

Wu SH, et al. (2019). The effects of afforestation on soil bacterial communities in temperate grassland are modulated by soil chemical properties. *PeerJ* 2019. https://doi.org/10 .7717/peerj.6147.

Xie WY, et al. (2018) Antibiotics and antibiotic resistance from animal manures to soil: a review. *European Journal of Soil Science* 69(1), 181-195. https://doi.org/10.1111/ ejss.12494

Xun W, et al. (2021) Specialized metabolic functions of keystone taxa sustain soil microbiome stability. *Microbiome* 9(1) https://doi.org/10.1186/s40168-020 -00985-9

Yin Y, et al. (2019) Soil microbial character response to plant community variation after grazing prohibition for 10 years in a Qinghai-Tibetan alpine meadow. In: *Plant Soil*, pp. 175–189. https://doi.org/10.1007/ s11104-019-04044-7.

Yuan X, et al. (2021) Development of fungal-mediated soil suppressiveness against fusarium wilt disease via plant residue manipulation. *Microbiome* 9(1) https://doi .org/10.1186/s40168-021-01133-7

Zanella PG, et al. (2021) Grazing intensity drives plant diversity but does not affect forage production in a natural grassland dominated by the tussock-forming grass

Andropogon lateralis Nees. *Scientific Reports* 11, 1–11. https://doi. org/10.1038/ s41598-021-96208-8.

Zhang FQ, et al. (2016) Dominance of ammonia-oxidizing archaea community induced by land use change from Masson pine to eucalypt plantation in subtropical China. *Applied Microbiology and Biotechnology* 100, 6859-6869. https://doi.org/10.1007 /s00253-016-7506-8

Zhao Z, et al. (2020) Fertilization changes soil microbiome functioning, especially phagotrophic protists. *Soil Biology and Biochemistry* 148, 107863. https://doi.org /10.1016/j.soilbio.2020.107863

Chapter 3

Implementing sustainable land use change programmes

Liz Lewis-Reddy, ADAS Policy and Economics, UK

1 Introduction

To meet the needs of a growing global population, reverse the decline in global biodiversity and adapt to the challenges of climate change, we are going to have to change the way we farm. One of the ways Governments across the UK have sought to address these challenges is through the creation of publicly funded agri-environment schemes (AES). Originally designed to have a singular focus on the protection and enhancement of on-farm biodiversity, modern iterations of these schemes have evolved to reflect the growing awareness of the need to promote a multifunctional landscape. This broader perspective allows for the recognition of the true value of these sustainably managed landscapes through the delivery of a diversity of outcomes; from food, to wildlife habitat, to carbon sequestration. This chapter explores how these schemes have evolved alongside the principles of sustainable land management (SLM), how they are shaped by the culture of the UK nations within which they develop, and how the role and importance of the landowner has changed from a passive recipient of instruction to an active participant in the design process. The chapter will also

http://dx.doi.org/10.19103/AS.2024.0134.18

explore a case study scheme whose ambition of economic viability achieved through ecologically resilient land management enables us to explore the challenges and opportunities associated with achieving SLM.

2 Sustainable land (SLM) management in a British context

To sustainably meet the needs of an estimated 9.1 billion people in 2050[1] (United Nations, 2017), global agricultural productivity (as measured by total factor productivity) would need to increase from the 2010–2018 average annual growth rate of 1.51% to a growth rate of 1.75% per annum (Steensland and Zeigler, 2021). Given limited natural resources and the continued degradation of resources already in use, increases in agricultural productivity will need to occur through gains in efficiency, intensification (e.g. increasing yield per hectare) or technological change. One example of such technological change is the widespread adoption of techniques associated with SLM.

The United Nations defines SLM as 'the use of land resources, including soils, water, animals and plants, for the production of goods to meet changing human needs, while simultaneously ensuring the long-term productive potential of these resources and the maintenance of their environmental functions'.[2] To deliver SLM, Sayer et al. (2012) proposed 'ten principles for a landscape approach to reconciling agriculture, conservation, and other competing land uses', a key principle of which is multifunctionality. The objective of multifunctionality can be achieved through the delivery of a range of outputs (environmental, economic, and social) from each hectare of land use, with each output delivered in such a way that it does not undermine the potential delivery of another. Agricultural production that delivers benefits for biodiversity while maintaining sustainable quantities of food produced without an over reliance on antimicrobial products (i.e. antibiotics and anthelmintics) would be one example of how a multifunctional approach could be applied to land management. The potential for multifunctionality underpins recent natural resource legislation enacted across the UK, with subtle variations reflecting the context (socioeconomic, environmental, and political) of each devolved nation.

In Northern Ireland (NI), a multifunctional approach is reflected in the launch of 'A Sustainable Agricultural Land Management Strategy for Northern Ireland'. In this example, there is a specific focus on working with the landowning community to enable behaviour change. Through better data and upskilling, the objective is to ensure that agricultural productivity is considered in parallel with environmental performance. Scotland has a slightly different focus as

---

1 https://www.unfpa.org/press/new-population-projections-underline-need-voluntary-family-planning-programmes.
2 https://www.fao.org/land-water/land/sustainable-land-management/en/.

illustrated through the various iterations of their 'Land Use Strategy'.[3] These strategies have a focus on transitioning to a low carbon economy through a greater understanding of the 'value' of Scotland's natural assets whilst enabling a more positive and constructive engagement between landowners and communities (as advocated through the Land Rights and Responsibility Statement (LRRS)).[4] In England, the 25 Year Environment Plan approach to SLM has a focus on improving the economic health of the agricultural sector through improvements in economic, ecological, and animal health resilience. The mechanisms that underpin this plan, such as the Sustainable Farming Incentive,[5] aim to reward farmers and land managers for the delivery of public goods.

The approach to implementing SLM in Wales embeds elements of all these approaches. This nation has considerable environmental assets (estimated to generate in excess of £8.8billion in goods and services annually (9% of GDP)) (National Trust, 2001), with most (88%)[6] managed under an agricultural system based in less favoured areas (LFA).[7] SLM has, understandably, been a focus of key legislation (Well-being of Future Generations Act, Environment (Wales) Act, and the Planning (Wales) Act). The Environment (Wales) Act, in particular, sets out nine principles or ways of working towards the objective of the sustainable management of natural resources. These principles include a focus on the following:

- Adaptive management: To ensure that the evidence gathered following the implementation of an action is reviewed and that any lessons learnt will be used to improve future actions and outputs;
- Spatial scale: Whereby any actions to repair and prevent serious damage to ecosystems are undertaken at an ecosystem scale;
- Collaboration and cooperation: Across and between all stakeholders within an ecosystem;
- Public participation and wider stakeholder engagement: To ensure buy-in to the concept of the value of the natural environment;
- Benefits and intrinsic value of ecosystems (species, habitats, and ecosystem services): These are identified and valued within decision making;
- Preventing damage to ecosystems: Through the recognition that the long-term sustainability of ecosystem service delivery relies on the restoration of healthy, diverse, and robust ecosystems; and

3 https://www.gov.scot/publications/scotlands-third-land-use-strategy-2021-2026-getting-best-land/.
4 https://www.gov.scot/publications/scottish-land-rights-responsibilities-statement/.
5 https://www.gov.uk/government/publications/sfi-handbook-for-the-sfi-2023-offer.
6 https://gov.wales/sites/default/files/publications/2019-06/agriculture-in-wales-evidence.pdf.
7 LFA refers to land located and included in the list of less favoured areas adopted by Article 2 of European Council Directive No.75/268EEC. In the UK, this includes land generally suitable for extensive livestock production and for the growing of crops for livestock feed, but agricultural production is restricted by soil, relief, aspect or climate conditions.

- Taking account of the short-, medium-, and long-term consequences of actions (including farming practices) to support the 'resilience of ecosystems'.

These principles underpin the development of Wales' post-Brexit agricultural support system, the Sustainable Farming Scheme. The same can be said for each of the aforementioned policies targeted at the delivery of SLM across the UK. Providing public funds to deliver land management focussed policy priorities is not a new approach. It is the principle that public funds could (and should) be used to pay for the environmental outcomes which arise from changes to farming practices that represent a shift in focus. Seventy years ago, with food rationing looming large in the collective British consciousness, the primacy of food production resulted in large swathes of land being given over to intensive single-use livestock or crop production. This approach became enshrined in various domestic (upland drainage policies) and EU policies (early version of the Common Agricultural Policy (CAP)), targeted at stimulating primary production. Over time, the narrow focus of these mechanisms and the negative impact they had on the natural environment, began to shift public opinion. Pressure began to mount on the UK Government to find a way for public funds to expand beyond subsidising food production and towards stewardship (Buckwell et al., 1997; Carey et al., 2002; Walker et al., 2004; Wynn, 2002). It was this change in approach that led to the development of the UK AES (Carey et al., 2002).

3   Evolution of the concept of SLM: from from landscape protection to ecosystem service delivery

In 1986, Countryside Stewardship initiatives, or AES, in the UK were formalised in law via the Agriculture Act (Buckwell et al., 1997). This led to the development of the Environmental Stewardship Act (ESA) (Carey et al., 2002; Coates, 1997) in 1987. The premise of the ESA was to 'protect the landscape, wildlife, and historic interest of specific areas within the UK that are of national environmental significance' (DEFRA, 2004).

In 1991, the ESA was expanded to incorporate the wider countryside through the creation of the Countryside Stewardship Scheme (CSS) (Carey et al., 2002; Coates, 1997). The CSS was taken over by the Ministry of Agriculture, Fisheries and Food (MAFF) in 1996, and had its remit expanded further in 2001 when MAFF became the Department for Environment, Fisheries and Rural Affairs (DEFRA). The multiple objectives of the new scheme came to include protection of all features of the rural landscape (DEFRA, 2004). These included biodiversity, wildlife habitats, archaeological sites, and historic features. Such changes in remit were enabled through reforms of the CAP (MacSharry Reform,

1992; Agenda 2000 reform; Luxembourg and Fischler Reform, 2003) into two pillars of payment. Pillar I, or *direct payments,* retained much of the original focus on public subsidy (i.e. were made on eligible land, conditional on land owners maintaining their land under good agricultural and environmental condition). Pillar II payments had a broader remit and were targeted at the *rural development* objectives of the individual members states and included novel agri-environmental payments and investment support.

This broadening of the remit reflected growing recognition (and an increased volume of evidence) of the importance of working beyond the boundaries of individual farms when attempting to address landscape-scale objectives. The Environment Stewardship Scheme in England 'at scale' introduced financial incentives or 'bonuses' to support coordinated action across landscapes (e.g. measures to improve water quality in a catchment such as fencing off watercourses). By affecting change at a landscape scale, the delivery of ecological outcomes became important to not only the public sector but also the private sector. With farmers working together to implement land management practices that benefited water quality, flood water management and beyond, those industries for whom these outcomes were a fundamental part of their business model began to take interest. Private sector initiatives often fell under the banner of Payment for Ecosystem Service (PES) schemes. PES are voluntary agreements between at least one 'seller' and one 'buyer' conditional on the supply of a well-defined ecosystem service, or a land use presumed to produce that service (Wunder, 2005).

Rather than being based on compensating landowners for enhancing the ecological value of farmland, these business-to-business schemes were based on the principle of paying for a defined environmental deliverable. These novel mechanisms represented a step change away from the early AES of the CAP. The environmental deliverable associated with a land management intervention was no longer a by-product of 'not farming properly'. The environmental outcome was now a primary commodity. Under a PES scheme, the economic resilience of the farm business was now linked to ecologically SLM.

4 Challenges and lessons from implementing SLM programmes

As these novel mechanisms became more established, they became a testing ground for more commercial approaches to addressing key challenges associated with publicly funded schemes (e.g. demonstration of value for money). One such approach was to set the payment rates for capital works and the delivery of environmental outcomes via 'Reverse Auctions'. Through this approach, farmers opt into a bidding process, offering to manage their land under prescribed management options at a price that more closely reflects the actual cost of doing so. To determine whether these approaches resulted in a more cost-efficient investment, Elliott et al. (2015) considered aspects of auction design and farmer responses to auctions through a combination of laboratory experiments, simulation modelling, and farmer workshops. The analysis demonstrated that over time, although the environmental outcomes of the schemes did not diminish, collusion between participants impacted the price at which interventions were set. This weakness in the reverse auction approach highlighted an opportunity for future scheme implementation; even within commercial environments, landowners can and will work together to achieve the best outcome for their collective businesses. The emphasis on the cost-effectiveness of schemes, however, remains a consideration for investors (public and private). For the public sector, demonstrating value for money is essential when associated with a limited pot of funding. For private sector investors who are participating in, or developing, schemes that are designed to achieve specific (and reportable) objectives, demonstrating delivery is key. The issue of 'deadweight' for both investors is therefore a significant problem.

Deadweight in an AES is defined as 'outcomes which would have occurred without intervention' (HM Treasury, 2003). Deadweight is important when estimating the net impact of AES on environmental delivery and wider behaviour change. However, the perception that it is a significant feature of publicly funded schemes may be overstated. An evaluation of an earlier Welsh AES, Tir Gofal (Agra Ceas, 2006), found a very small percentage (e.g. 5%) of deadweight within the scheme. Only 5% of those enrolled in the scheme would have undertaken the activity irrespective of funding. However, even for these 5%, the additional funding significantly increased the amount of work undertaken (i.e. less work would have been done and without support it may have been deferred to the future). A similar outcome was observed from an evaluation of an early Higher Level Stewardship (HLS) scheme in England. CCRI (2010) determined that in the absence of HLS scheme payments, 79% of the environmentally focussed management work would not have been undertaken, equating to only 21% deadweight.

Modelling work (Uthes et al., 2010; Kuhn et al., 2019) has suggested that the low levels of deadweight observed may reflect the environmental context within which participating farms are operating. Farms participating in AES were more likely to be able to make money from participation than other agricultural businesses. This could be because the management practices already underway on farm were similar to those dictated by the AES. The outcome of this would be on-farm costs which were well below the scheme payment rate (Uthes et al., 2010). In essence, farms which select interventions that match their environmental context and capability do not represent deadweight but are instead illustrating the opportunity (economic and environmental) associated with appropriate targeting of land management interventions.

Post Brexit, the agricultural support framework across the UK has shifted away from the two pillar segregation of agricultural production from environmental deliverables. The new approaches, as outlined earlier in this chapter, are targeted at supporting SLM, and have ambitious objectives, including the delivery of sustainable food production; to provide a response to the climate emergency; and to halt, and ultimately reverse, the loss of biodiversity. These objectives will require working at scale across many land holdings. Successful implementation of SLM across landscapes will require high levels of uptake and collaboration from across the landowning community. The main factors that will impact uptake are linked to the three overarching pillars of sustainability: the economics of the farm business, the environmental constraints and opportunities within which these businesses operate, and the capacity and capability of the farmer, and the wider landowning community.

4.1 Economic context

Agricultural production is subject to a wide variety of external variables (e.g. price fluctuations and climate stresses). Diversification (agricultural and non-agricultural) of the farming business to include a variety of income streams has often been promoted as a means by which can farms can reduce their reliance on, at times, an unreliable source of income. Previous AES have sometimes been viewed as one such means of diversification but research on the economic outcomes of AES (the best proxy for understanding the potential profitability of some aspects of SLM) is mixed. Sauer et al. (2012) noted that farms participating in voluntary AES tended to become less specialised and more diversified with respect to their production structure. However, the 'costs plus income foregone' model for the setting of AES payment rates is predicated on the presumption that participating farmers will not profit from the implementation of AES measures. This objective is borne out by the fact that there is a general consensus within the farming community that implementing measures to benefit the environment or biodiversity incurs a direct economic cost through a reduction in primary production capability. For the implementation of sustainable management schemes, this perspective will have to be overcome with robust economic evidence. For example, Harkness et al. (2021) determined that for some farming systems (e.g. dairy, general cropping, and mixed farms), participation in AES (as an indication of system and business flexibility) can have a demonstrably positive impact upon long term financial stability. The authors attributed the increased stability not only to the value of the payment received but also the potential of these farming practices to increase natural processes on farm alongside an associated reductions in input cost (Harkness et al., 2021).

By contrast, Udagawa et al. (2014) observed no evidence that entry into AES (Entry Level Stewardship (ELS)) in England significantly increased levels of farm income and found that there was a modest level of cost for participation, at least initially. The authors suggested that earlier studies which suggested otherwise (Harrison and Jones, 2010) were too focussed on a subset of farms and therefore not representative of the wider sector. A causal relationship between AES payment level and financial stability was also disputed by Zhu and Oude Lansink (2012). The authors identified an inverse relationship between the share of total subsidies in total farm revenues and business efficiency, termed an 'income and insurance effect'. They demonstrated that farm business models which obtain the majority of their income from subsidies or public programmes have less motivation to improve efficiency.

It is important, however, to understand what is driving this variation in response to subsidy across farm types and environmental contexts. The outcome of these studies suggests that a degree of self-selection is taking place whereby the profitability of participation in AES is not solely a reflection of payment rates but also of the environmental attributes of the participating farms.

4.2 Environmental constraints and opportunities

By contrast with the findings from dairy, general cropping and mixed farms, Harkness et al. (2021) found that AES payments on LFA farms reduced the stability of income, despite these farms receiving more money from agri-environment schemes per hectare, on average, than other farm types. This counterintuitive outcome was linked to the fact that LFA farms are based in agriculturally challenging environments. As such, there would be little room for improvements in productivity through the implementation of sustainable management. Soil management-focussed AES options may also generate improvements in yield whereas the same options applied to upland soils are unlikely to yield a similar productivity enhancement. Upland areas are, however, capable of delivering significant environmental outcomes (e.g. water quality and quantity management, and carbon sequestration). The outcome of the Harkness et al. (2021) study could be a reflection of the misalignment between outcomes and payment rates. In essence, by only valuing the outcomes related to improvements in soil structure the additional environmental delivery achieved by these LFA farms was not being accounted for in the payment rate.

By contrast with LFA farmers, Harrison and Jones (2010) found that farms with an abundance of features (e.g. hedgerows, small field parcels) were able to profit from participation in the English entry-level scheme (ELS) for Countryside Stewardship through careful selection of measures. By selecting, intentionally or sub-consciously, options which would have limited income-foregone impacts, these farmers were able to maximise profitability. Farm businesses based in more featureless landscapes had to instead resort to more costly interventions which limited the overall profitability of each measure. It is worth noting that this profitability was viewed solely from the perspective of the payment rate and did not take account of any increase in natural processes arising from the intervention (as highlighted in Harkness et al., 2021). It could be argued that (as per the previous LFA farm examples) overall scheme design was flawed. If payment rate had been targeted at environmental output rather than capital delivery, the environmental output on both farm types would have been appropriately remunerated. The feature-rich farms would not have had to incur significant capital costs as they were already delivering value (albeit at lower rate of change), whereas increasing feature richness on the feature poor farms would have resulted in significant additional environmental value. This challenge of only 'paying for change' links back to the concerns over 'deadweight'. The aforementioned evaluation work undertaken by Agra Ceas (2006) and CCRI (2010), however, has demonstrated that deadweight is more perception than reality. There is always more than can be achieved, especially when landowners collaborate.

Lobley and Butler (2010), Morris et al. (2017) and Uthes et al. (2010) have flagged that multifunctional objectives also require multifunctional policy frameworks. Changes to agricultural support mechanisms to focus on SLM will need to be prefaced by in-depth policy analysis (and the delivery of a range of targeted interventions to increase capacity and capability of the landowning community) if contradictory effects are to avoided. Otherwise less agriculturally productive regions with their deminished financial capital, may struggle to respond to policy and be left further behind (Severini and Tantari, 2013).

4.3 Human capital

Work undertaken by Wynne-Jones (2013) highlighted that most farmers see their primary role as being producers of food, and few understand the concept of being producers of ecosystem services or other environmental outcomes. Those who participate in agri-environment schemes generally do so as a way to support an approach to farming that makes less demand of their time and resources, or as a risk management solution to smooth incomes during uncertain economic times (Ingram et al., 2013). This 'business as usual' option selection often reflects a farmer's understanding of what works best for their business but not necessarily an understanding of what the scheme is trying to achieve (Turner et al., 2013) and how it would benefit their business in the long term.

5 Building capacity and capability within the farm business

The ability of the farming community to respond to change is not simply a reflection of the current current farmer but is also dependent on the next generation. The presence of a successor has long been identified as a key driver of innovation and expansion of farm businesses (Barnes et al., 2014; Chiswell, 2014; Potter and Lobley, 1992). Farms with an identified successor and an established plan for the handover of responsibility tend to be motivated, as well as increasingly disposed to adaptation, investment, and expansion (Chiswell, 2014). This greater receptiveness to innovation was highlighted by Vecchio et al. (2022) in their work on the adoption of precision agriculture (PA) by Italian farmers. The authors identified three groups of farmers: adopters, non-adopters, and planners. Adopters were typically young, university educated farmers who frequently informed themselves on new innovations by reading relevant literature.

Busse et al. (2014) and Läpple and Hennessy et al. (2015) also highlighted the importance of intergenerational factors influencing adoption. Both publications identified that younger, highly qualified farmers, who had been given management responsibility were more likely to be open minded to innovation. Failing to embed collaboration at all stages of innovation and scheme design can result in missed opportunities to engage farmers who could be classed as 'early adopters'. Work undertaken by Läpple and Hennessy (2015), Toma et al. (2018), and Farrell et al. (2022) identified that public policy interventions which promoted structural change and earlier inter-generational transfer may facilitate adoption. In their study on the Irish organic farming sector as part of the Irish EIP-Agri funding programme, Läpple et al. (2015) found that targeted policy interventions that attract the next generation back to farming have the potential to enhance innovation acceptance (embracing novel opportunities).

The diversity of perspectives within a farm business in the wider farming community will consequently require a diverse range of mechanisms to support change. Knowledge exchange service providers associated with the agricultural community have responded to this by not only providing a range of different methods for promoting knowledge exchange activities, but also bundling methods (e.g. small group learning and coaching) to best suit different types of farmers and or different types of problems. Research undertaken by Nettle et al. (2022) reported that the best methods (based on extent, reach, and time to change) were small group learning and one-to-one advice or coaching. They noted that a focus on farmer needs, and the 'journey of change' were key to having a greater impact. Supporting farmers on this journey of change towards more SLM will also require improved links between the farming community and the development of the evidence base for change. As these developments often take place through institutions that seem remote from on-farm practicalities, this can represent a challenge to gaining uptake across the farming sector.

6 Building capacity and capability around the farm business

In their paper exploring the on-farm experimentation (OFE) process, Lacoste et al. (2022) reiterated the importance of restructuring the farmer-researcher relationship. The OFE process refers to a collaborative experimental research process that brings agricultural stakeholders together to develop and implement mutually beneficial experiments. OFE is based on three principles: (1) experimentation occurs on-farm; (2) interests and insight from all participants are explicitly acknowledged at the outset, and (3) experimentation is a deliberative, iterative process of joint exploration. This helps to build trust not only between the stakeholders taking part in the OFE but in the outcome of the research.

The importance of building trust is reinforced by work undertaken by Ketterings (2014). Their research suggested that a successful knowledge transfer programme is one where the actual message might not be as important as the trust-based partnership and understanding of farmer reality that leads to asking the right questions. Based on the nutrient management experiences of a dairy farm in New York State, Ketterings (2014) listed the key ingredients for farm level adoption as: (1) understanding the concerns and recognition that change is necessary among all involved; (2) identification of win-win situations first; (3) posing of relevant questions, and generation of believable results based on reliable data (replicated trials where feasible); (4) farmer involvement and accountability in the process (farmers as drivers of the adaptive management process), and (5) development and maintenance of a trust-based relationship between farmer, farm advisor, and university researcher. Ketterings (2014) identified that an innovative, outcome-focussed adaptive management approach that includes research, extension, and a focus on human dimensions (a people-based approach) was the most effective way to obtain long term on-farm change.

Their analysis of dairy farm experiences included a review of three different adaptive management approaches, implemented at field, whole farm, and regional/state levels. In the study, the participant farmers, in collaboration with the research institutions (Cornell University), played a fundamental role in the development, design, and implementation of on-farm trials. In addition, the outcome of the trials was evaluated through the production of summary reports and discussed at crop management team meetings. This enabled context-specific conclusions and impact to be reported and acted upon. The lessons learned from these on-farm research partnerships were then explored with farmers across the wider catchment, the outcome of which was significant improvements in field, farm, and regional/state level balances for nitrogen and phosphorus in New York State.

Research undertaken in Germany by Busse et al. (2014), which focussed on the German Agricultural Innovation System, reinforced the importance of the involvement of farmers at the design stage of innovation. This study illustrated that the role of farmers in generating innovations is often not recognised by researchers. The outcome of this lack of recognition is the inability of the farmers to function not only as adopters but also co-designers or co-developers. A review of 11 case studies from the international RETHINK research programme by Šūmane et al. (2018) identified the importance of recognising all stakeholders as equal co-authors of knowledge generation and facilitators of knowledge exchange. They also recognised that national level ambitions are best delivered by supporting local-level initiatives; including those linked to regulation.

This latter point was demonstrated two case studies, one involving a group of organic Austrian farmers, and another a group of smallholders in Latvia (Šūmane et al., 2018). The former regarded inflexibility around dates for spreading manure and slurry as not enabling farmers to take into consideration regional weather. The latter was constrained by national regulations related to the administration of vaccinations. In both cases, by not being part of the decision-making process, the farmers felt that their expertise and experience-based knowledge and skills were being ignored. The impact of their exclusion ultimately undermined the sustainability of their agricultural practices and their ability to drive innovation within their own systems. The paper by Nettle et al. (2022) previously referred to, outlined the importance of locally designed innovations, developed through trusted networks of farmers, scientists, and commercial firms, with specific reference to the Australian dairy sector. The flexibility to iterate and adapt at a local level proved to be the most successful strategy for increasing uptake across the sector.

The process of a farmer adopting a new technology or practice has been conceptualised as a 'dynamic learning process', which is influenced not only by the characteristics of the farmer (Hill, 2007) and the innovation, but also by the broader social environment (Prager et al., 2017). A farmer's social and information networks are important in becoming aware of a problem, evaluating known information, and shaping the decision whether (or not) to adopt. It is influenced by farmer-to-farmer interaction, and exchange of information between peers and with specialists (Prager et al., 2017). Whether production or sustainability focussed, the research exploring the influence of scheme design on farmers' adoption of new practices or technologies/innovations identifies similar influencing factors. These include the role of the farmer in idea development and implementation and the establishment of supportive structures (policy and institutions, including centres of excellence) which facilitate collaboration and innovation development at a local level (Läpple et al., 2015).

6.1 Role of Institutions

Reflecting on the implementation and institutionalisation of Innovation Platforms and climate smart agriculture in sub-Saharan Africa, Schut et al. (2016) and Makate (2019) demonstrate how adaptation of these interventions from their 'old' institutional way of working, towards approaches where stakeholders have collective agency, may increase impact. Schut et al. (2016) identified that the efficacy (and impact) of the Innovation Platform is as much a result of the quality of the multi-stakeholder process (jointly identifying constraints, sharing information, and learning) as it is of the specific innovation it develops.

Schut et al. (2016), however, recognised that many Innovation Platforms are structured along disciplinary lines, have a long history of 'top down' technology development and transfer, and have limited expertise with regards to institutional innovation, facilitation of interactive multi-stakeholder processes or how to address structural power inequalities between actors in the supply chain. Effective dissemination (adoption, use, uptake or commercialisation of existing knowledge) of alternative approaches to land management requires an understanding of local farming systems and contexts, which can only be achieved through strong linkages and active participation of a range of actors. Makate (2019) reinforced this observation by highlighting the value of resource pooling (capital, knowledge, technology, land, etc.) and that interventions work better when multiple stakeholders are harnessed as a resource and brought to the table.

In the Welsh context, the development of robust institutions supporting Innovation Platforms is essential to supporting change. Statistics gathered by Defra in 2015[8] highlight that the agricultural workforce in Wales comprises approximately 2% of the Welsh population. Despite their responsibility for the management of over 70% of the Welsh land area, this workforce is not required to achieve any educational training beyond secondary education. Only 20% of the workforce has a qualification which equates to level 4 (equivalent to a Foundation Degree or Higher National Diploma) or above, compared to 40% in other sectors of the economy (Jones, 2015). It could be argued that this lack of education explains the reliance of the Welsh agricultural sector on anecdotal approaches to farming and ecological awareness.

---

8 https://www.gov.uk/government/statistical-data-sets/structure-of-the-agricultural-industry-in-england-and-the-uk-at-june

The above insight from the literature reflects the importance of considering economic, environmental and social capital when attempting to implement change at all levels, from farm business to sector wide. Successful implementation of SLM will therefore require:

- appropriate targeting of locally relevant multifunctional options, rather than development of broad 'one size fits all' approaches, to ensure the best outcomes are achieved for both the farm business, the investor, and wider society;
- close collaboration and communication between policy makers, researchers, landowners advisors and the wider community at all stages of scheme development and implementation to ensure higher levels of uptake, local community buy-in and better biodiversity outcomes at all scales;
- implementation of verifiable outcome-based performance assessments tailored to the environmental context of each farm business to embed flexibility and adaptive management, whilst giving landowners the confidence and expertise to engage; and
- financial remuneration which fairly values of the output and cost of implementation.

7 Case study: Pumlumon Project

The Pumlumon Project (PP) is a place-based project named after the Pumlumon Mountain (Fig. 1). It covers a watershed area of 40000 ha, contains the highest part of the Cambrian Mountain range and straddles the counties of Powys

Figure 1 Landscape in the Pumlumon project area.

and Ceredigion. The whole project area is home to 15000 people, spread across 11 local communities. There are 250 farms in the project area with farming, forestry, and tourism the main economic activities. It is also the largest watershed in Wales with the reservoirs and streams that drain into the Wye, Severn, Rheidol, Dyfi and Leri river catchments supplying water to four million people in England. Encompassing over 9000 ha of key habitats including river valleys, semi-natural woodland, species-rich grassland, heather moorland, and blanket bog, at the heart of the project area sits the 5000 ha Pumlumon Site of Special Scientific Interest (SSSI), currently in unfavourable ecological condition and declining.[9]

For over 10 years, the Montgomeryshire Wildlife Trust (MWT) has been working with local communities, land managers, statutory agencies, and both local and national businesses, to restore and enhance the resilience of the ecosystem within the project area; piloting an integrated approach whereby the ecosystem services (i.e. water quality, flood risk reduction, carbon safeguarding) can be better delivered via the mechanism of SLM.

7.1 Economic

Early on in project development, MWT recognised the importance of exploring novel and innovative economic models that could help support the long-term sustainability of the rural economy. The degree of scientific uncertainty around verifiable delivery of ecosystem services represented an early challenge to secure investment. However, this was overcome in the short term by working closely with the research community to develop robust methods for quantifying change. The project team also undertook extensive research with the local and national business community to better align the supply of ecosystems services from the project area with private sector demand.

In 2014, substantial capital funding was provided from the National Lottery for a major ecotourism infrastructure build within the project area, the Dyfi Osprey Project. This development had a significant impact on the local community (valued at £612500/year)[10] and highlighted the economic opportunities associated with nature-based tourism.

7.2 Environmental

In collaboration with the landowning community, the project team undertook a thorough review of environmental deliverables from the project area (i.e. what can, or could be, produced in the Pumlumon project area and could contribute

9 http://wyndrushwild.co.uk/wp-content/uploads/2012/11/MSE_Survey-of-Summit-Vegetation-on-Pumlumon-SSSI
    -contract-science-report.pdf
10 Dyfi Osprey Project Economic Impact Summary (2010–2013)

to the area's future income and sustainability (carbon sequestered, sustainable food, water quality, and quantity management). This included a spatial mapping element and quantitative element. The quantitative element, where reasonable, used economic methods to estimate potential monetary value associated with each deliverable.

The mapping of key environmental deliverables enabled a detailed assessment of the spatial variation in the supply. This 'opportunity mapping' (Fig. 2) also highlighted where interventions could be most appropriately targeted to deliver maximum outcomes.

7.3 Social

From the project's inception through to implementation, a farmer/landowner working group was established to co-produce each of the project. Early opportunities were explored with 'early adopters' from the group who became project champions within the wider community. The trusted relationships that developed because of this approach continue today and have enabled the exploration of new ideas from PES to more locally based collaborative economies.

Since its inception in 2008, the PP continues to grow. The farmer/ landowner working group remains active and continues to collaborate on

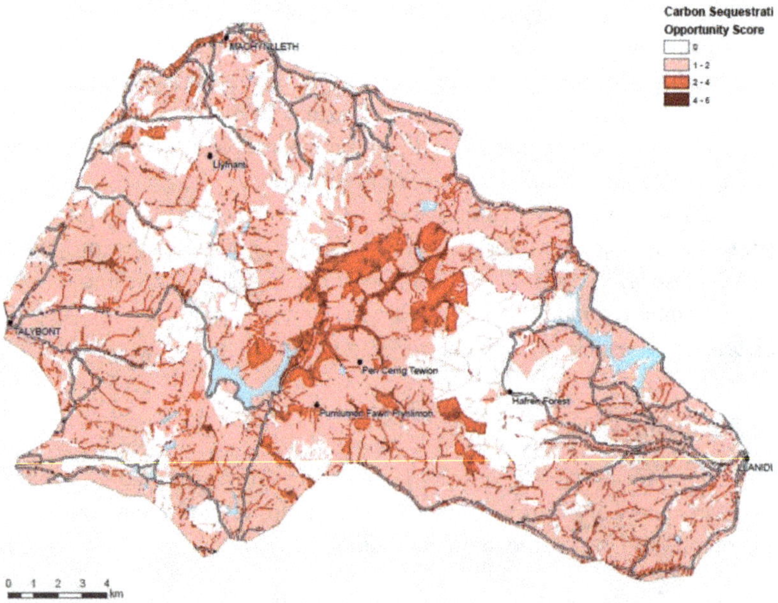

Figure 2 Opportunity mapping of peatland restoration opportunities in the Pumlumon project area. Darker areas represent highest restoration priority.

emerging opportunities with MWT (Fig. 3). A number of the lessons learnt via the project (e.g. the importance of collaboration and co-production with local communities, the ability to embed private sector investment mechanisms into farm business models and, most importantly, the capacity to produce a range of viable ecosystems services through the application of SLM (Fig. 4a and b)) helped to underpin the SLM objectives of the current Welsh legislative framework.

8   What the future could look like

SLM models such as those explored in the Pumlumon Project demonstrate that a future where the decline in UK biodiversity is halted, recovery is well underway, and the direct links between the health of the natural environment and the health and well-being of society are valued and understood are possible (Fig. 5). To achieve this, the UK requires an economically viable landowning community working with the environment to ensure that it can provide both for the needs of the natural environment and society.

The post-Second World War decoupling of ecosystem processes from agricultural practice damaged the natural environment and degraded its ability to deliver ecosystem services (e.g. water quality and quantity). Valuing agricultural fields solely by crop yield meant that the collateral damage to the wider environment became irrelevant. Value should instead incorporate the multiple societal benefits accrued from SLM (e.g. clean water, reduced flood

Figure 3 Pumlumon Project knowledge transfer event with the International Federation of Agricultural Journalists and Landowner working group.

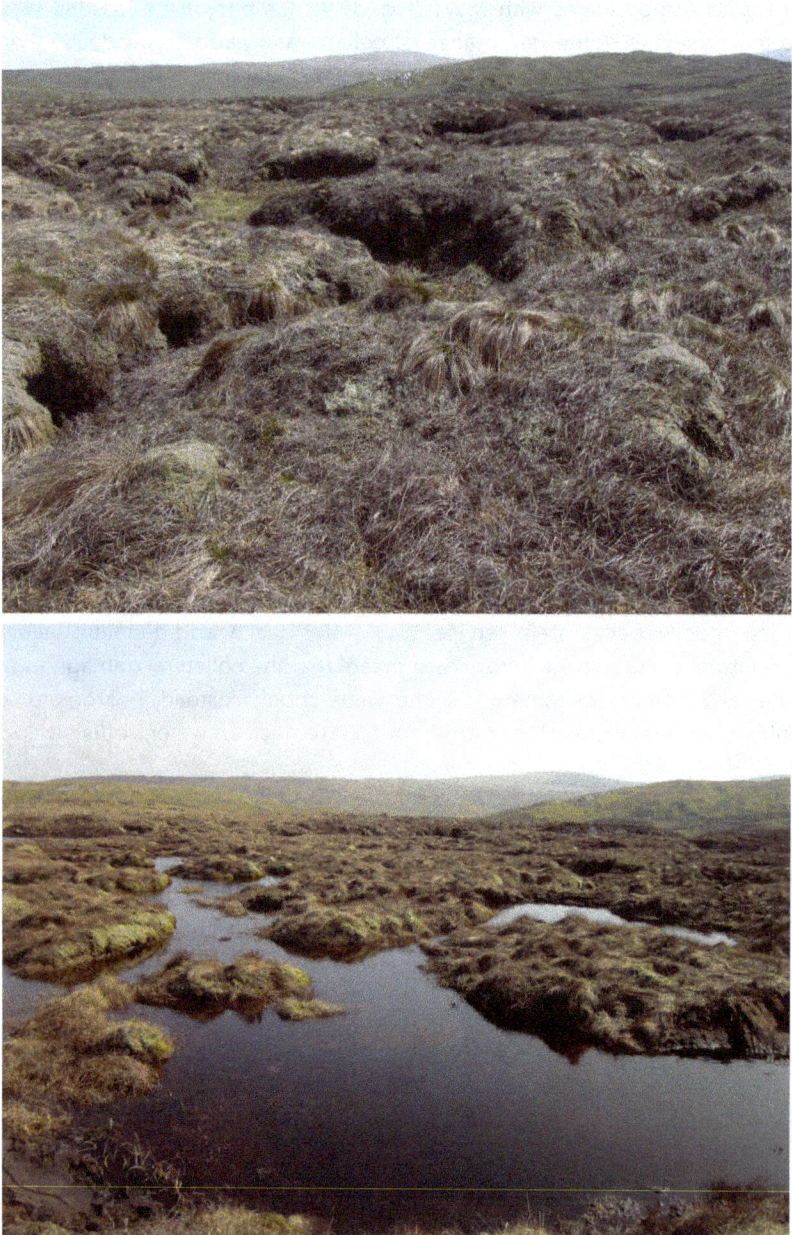

Figure 4 (a) Degraded peat bog before restoration of water table. (b) Degraded peat bog after restoration of water table.

Figure 5 Restored blanket bog within the Pumlumon project area. The image is reproduced with the permission of the Montgomeryshire Wildlife Trust.

risk, reduced greenhouse gas emissions) as well as the true 'costs' associated with production. If SLM is not pursued, unsustainable practices which negatively impact upon the delivery of the associated benefits will ultimately reduce the overall value of what is being produced, shrinking the profit margin and resulting in economic instability. It is damaging to silo one area to production and another to environmental deliverables.

Deploying a single scale of merit (e.g. number of plant species within a given area) should also be avoided. Biodiversity across a range of farm types and elevations (e.g. lowlands and uplands) is different, but equally valuable. The implementation of SLM needs to avoid being reductionist, and to exercise caution in the use of proxies for simple indices of quality or change. Effort needs to be expended in understanding local environments and tailoring solutions and monitoring programmes. The on-going reduction of expert input, poor monitoring, lack of standardised and accepted metrics, delivery of one-size-fits-all solutions, and single outcome approaches have all contributed to poor outcomes. The objectives of the integrated natural resource policies proposed across the UK suggest that the recovery of nature and production of food is compatible, and multiple outcomes are possible. To be successful in the long term, however, they will need to embed the following key principles:

- Create multiple outcomes – ecosystem services produced everywhere and anywhere can make a contribution to biodiversity;
- Use societal funds for societal good – public funds should be used to provide equitable access to health and well-being benefits, reductions in flood risk, improvements in water and air quality, and safeguarding of vulnerable habitats and species;
- Foster greater public understanding of the value of the natural environment and role of consumer choice; and
- Support institutions capable of upskilling and professionalisation of those who work with the natural environment.

If these principles are adhered to, the recovery of biodiversity will be secured and the economically and ecologically sustainable management of the UK environment into the future will be assured.

9   Conclusion and future trends

Significant work is underway across the devolved nations of the UK, and the wider world, to better quantify and value the ecosystem services arising from SLM. Further research is required to ensure these metrics are robust enough to give the private sector the confidence to invest, sensitive enough to detect change (even over short timescales) to warrant public investment, and resilient enough to help guide decision making into the future. ADAS is working closely with the devolved UK nations on each of the SLM programmes. We are helping them to develop structures and tools (earth observation, artificial intelligence, natural capital accounting and governance protocols) to help address these needs, but there is always more work to be done.

10   Where to look for further information

UK Sustainable Land Management Policy documents can be found here:
Northern Ireland: SLM strategy

- https://www.daera-ni.gov.uk/sites/default/files/publications/daera/16 .17.079%20Sustainable%20Land%20Management%20Strategy%20final %20amended.PDF.

Scotland: Land Use Strategy

- https://www.gov.scot/publications/scotlands-third-land-use-strategy -2021-2026-getting-best-land/.

Scotland: Land Rights and Responsibilities

- https://www.gov.scot/publications/scottish-land-rights-responsibilities -statement/.

England: Environmental Improvement Plan

- https://www.gov.uk/government/publications/environmental -improvement-plan.

Wales: Sustainable Management of Natural Resources

- https://naturalresources.wales/media/678063/introducing-smnr-booklet -english-final.pdf.

11   References

Agra CEAS Consulting (2006). Socio-economic Evaluation of TIR GOFAL. Final Report for Countryside Council for Wales, Fforwm Tirlun and the Welsh Development Agency, p. 103.

Barnes, A., Toma, L., Mathews, K., Sutherland, L. and Thomson, S. (2014). Intensify, diversify, opt-out: testing farmer stated intentions to past and future CAP reform scenarios. Contributed Paper prepared for presentation at the 88th Annual Conference of the Agricultural Economics Society, AgroParisTech, Paris, France.

Buckwell, A., Blom, J., Commins, P., Hervieu, B., Hoferither, M., Meyer, H., von Rabinowicz, E., Sotte, E. and Sumpsi Vina, J. (1997). *Towards a Common Agricultural and Rural Policy for Europe*. European Commission, Brussels.

Busse, M., Doernberg, A., Siebert, R., Kuntosch, A., Schwerdtner, W., König, B. and Bokelmann, W. (2014). Innovation mechanisms in German precision farming. *Precision Agriculture* 15(4), 403–426. https://doi.org/10.1007/s11119-013-9337-2.

Carey, P. D., Barnett, C. L., Greenslade, P. D., Hulmes, S., Garbutt, R. A., Warman, E. A., Myhill, D., Scott, R. J., Smart, S. M., Manchester, S. J., Robinson, J., Walker, K. J., Howard, D. C. and Firbank, L. G. (2002). A comparison of the ecological quality of land between an English agri-environment scheme and the countryside as a whole. *Biological Conservation* 108(2), 183–197.

Chiswell, H. M. (2014). The importance of next generation farmers: a conceptual framework to bring the potential successor into focus. *Geography Compass* 8(5), 300–312.

Coates, D. (1997). UK policy for the ESAs. Grassland management in the 'environmentally sensitive areas'. In: Occasional Symposium No. 32, British Grassland Society, Lancaster.

DEFRA (2004, 26/08/2006). History.. Available at: http://www.defra.gov.uk/erdp/docs/css_esas_report/history.htm.

Elliott, J., Day, B., Jones, G., Binner, A. R., Smith, G., Skirvin, D., Boatman, N. D. and Tweedie, F. (2015). Scoping the strengths and weaknesses of different auction and PES mechanisms for Countryside Stewardship. Defra Project LM0105. Final report.

4

Farrell, M., Murtagh, A., Weir, L., Conway, S. F., McDonagh, J. and Mahon, M. (2022). Irish organics, innovation and farm collaboration: a pathway to farm viability and generational renewal. *Sustainability* 14(1). https://doi.org/10.3390/su14010093.

Harkness, C., Francisco, A., Semenov, M., Senapati, N., Shield, I. and Bishop, J. (2021). Stability of farm income: the role of agricultural diversity and agri-environment scheme payments. *Agricultural Systems* 187, 1-13.

Harrison, G. and Jones, J. (2010). A farm level assessment of the profitability of Entry Level Scheme participation in the Lincolnshire Wolds. 84th Annual Conference, Agricultural Economics Society, March 29–31.

Hill, B. (2007). *Business Competence*. Department of the Environment, Food and Rural Affairs.

Ingram, J., Gaskell, P., Mills, J. and Short, C. (2013). Incorporating agri-environment schemes into farm development pathways: a temporal analysis of farmer motivations. *Land Use Policy* 31, 267–279.

Jones, W. (2015). Independent Review of Learning delivered by FE colleges and the relevance of that delivery supporting Farm businesses in Wales.

Ketterings, Q. M. (2014). Extension and knowledge transfer: adaptive management approaches for timely impact. *The Journal of Agricultural Science* 152(S1), 57–64. https://doi.org/10.1017/S002185961300066X.

Kuhn, T., Schäfer, D., Holm-Müller, K. and Britz, W. (2019). On-farm compliance costs with the EU-nitrates directive: a modelling approach for specialized livestock production in northwest Germany. *Agricultural Systems* 173, 233–243.

Lacoste, M., Cook, S., McNee, M., Gale, D., Ingram, J., Bellon-Maurel, V., MacMillan, T., Sylvester-Bradley, R., Kindred, D., Bramley, R., Tremblay, N., Longchamps, L., Thompson, L., Ruiz, J., García, F. O., Maxwell, B., Griffin, T., Oberthür, T., Huyghe, C., Zhang, W., McNamara, J. and Hall, A. (2022). On-farm experimentation to transform global agriculture. *Nature Food* 3(1), 11–18. https://doi.org/10.1038/s43016-021-00424-4.

Läpple, D. and Hennessy, T. (2015). Exploring the role of incentives in agricultural extension programs. *Applied Economic Perspectives and Policy* 37(3), 403–417. https://doi.org/10.1093/aepp/ppu037.

Läpple, D., Renwick, A. and Thorne, F. (2015). Measuring and understanding the drivers of agricultural innovation: evidence from Ireland. *Food Policy* 51, 1–8. https://doi.org/10.1016/J.FOODPOL.2014.11.003.

Lobley, M. and Butler, A. (2010). The impact of CAP reform on farmers' plans for the future: some evidence from South West England. *Food Policy* 35(4), 341–348.

Makate, C. (2019). Effective scaling of climate smart agriculture innovations in African smallholder agriculture: a review of approaches, policy and institutional strategy needs. *Environmental Science and Policy* 96, 37–51. https://doi.org/10.1016/J.ENVSCI.2019.01.014.

Mills, J., Courtney, P., Gaskell, P., Reed, M. and Ingram, J. (2010). Estimating the incidental socio-economic benefits of Environmental Stewardship Schemes. Final Report by CCRI.

Morris, W., Henley, A. and Dowell, D. (2017). Farm diversification, entrepreneurship and technology adoption: analysis of upland farmers in Wales. *Journal of Rural Studies* 53, 132–143.

National Trust (2001). Valuing our environment: economic impact of the environment of Wales. *The Valuing Our Environment Partnership*, July 2001.

Nettle, R., Major, J., Turner, L. and Harris, J. (2022). Selecting methods of agricultural extension to support diverse adoption pathways: a review and case studies. *Animal Production Science*. https://doi.org/10.1071/AN22329.

Potter, C. and Lobley, M. (1992). Aging and succession on family farms: the impact on decision making and land use. *Sociologia Ruralis* 32(2/3), 317–331.

Prager, K., Creaney, R. and Lorenzo-Arribas, A. (2017). Criteria for a system level evaluation of farm advisory services. *Land Use Policy* 61, 86–98. https://doi.org/10.1016/J.LANDUSEPOL.2016.11.003.

Sauer, J., Walsh, J. and Zilberman, D. (2012). Producer behaviour and agri-environmental policies: a directional distance based matching approach. Agricultural and Applied Economics Association Conferences. Annual Meeting, Seattle, Washington, DC.

Sayer, J., Sunderland, T. and Gazoul, J. (2012). Ten principles for a landscape approach to reconciling agriculture, conservation, and other competing land uses. *Proceedings of the National Academy of Sciences of the United States of America* 110(21), 8345–8348.

Schut, M., Klerkx, L., Sartas, M., Lamers, D., Campbell, M. M., Ogbonna, I., Kaushik, P., Atta-krah, K. and Leeuwis, C. (2016). Innovation platforms: experiences with their institutional embedding in agricultural research for development. *Experimental Agriculture* 52(4), 537–561. https://doi.org/10.1017/S001447971500023X.

Severini, S. and Tantari, A. (2013). The effect of the EU farm payments policy and its recent reform on farm income inequality. *Journal of Policy Modeling* 35(2), 212–227.

Steensland, A. and Zeigler, M. (2021). Productivity in agriculture for a sustainable future. In: Campos, H. (Ed.). *The Innovation Revolution in Agriculture*. Springer, Cham. https://doi.org/10.1007/978-3-030-50991-0_2.

Šūmane, S., Kunda, I., Knickel, K., Strauss, A., Tisenkopfs, T., Rios, Id. I., Rivera, M., Chebach, T. and Ashkenazy, A. (2018). 'Local and farmers' knowledge matters! How integrating informal and formal knowledge enhances sustainable and resilient agriculture. *Journal of Rural Studies* 59, 232–241. https://doi.org/10.1016/J.JRURSTUD.2017.01.020.

Toma, L., Barnes, A. P., Sutherland, L.-A., Thomson, S., Burnett, F. and Mathews, K. (2018). Impact of information transfer on farmers' uptake of innovative crop technologies: a structural equation model applied to survey data. *The Journal of Technology Transfer* 43(4), 864–881. https://doi.org/10.1007/s10961-016-9520-5.

Turner, A., Wilson, L., Edgington, P., Nias, I. and Drake, B. (2013). Final report on project 23768 resource protection monitoring of uptake and management of ES options to address DWPA. Report for Natural England.

Udagawa, C., Hodge, I. and Reader, M. (2014). Farm level costs of agri-environment measures: the impact of entry level stewardship on cereal farm incomes. *Journal of Agricultural Economics* 65(1), 212–233.

Uthes, S., Sattler, C., Zander, P., Piorr, A., Matzdorf, B., Damgaard, M., Sahrbacher, A., Schuler, J., Kjeldsen, C., Heinrich, U. and Fischer, H. (2010). Modeling a farm population to estimate on-farm compliance costs and environmental effects of a grassland extensification scheme at the regional scale. *Agricultural Systems* 103(5), 282–293.

Vecchio, Y., De Rosa, M., Pauselli, G., Masi, M. and Adinolfi, F. (2022). The leading role of perception: the FACOPA model to comprehend innovation adoption. *Agricultural and Food Economics* 10(1), 5. https://doi.org/10.1186/s40100-022-00211-0.

Walker, K. J., Stevens, P. A., Stevems, D. P., Mountford, J. O., Manchester, S. J. and Pywell, R. F. (2004). The restoration and re-creation of species-rich lowland grassland on land formerly managed for intensive agriculture in the UK. *Biological Conservation* 119(1), 1–18.

Wunder, S. (2005). *Payments for Environmental Services: Some Nuts and Bolts*. Centre for International Forestry Research, Indonesia.

Wynn, G. (2002). The cost-effectiveness of biodiversity management: a comparison of farm types in extensively farmed areas of Scotland. *Journal of Environmental Planning and Management* 45(6), 827–840.

Wynne-Jones, S. (2013). Ecosystem service delivery in Wales: evaluating farmers' engagement and willingness to participate. *Journal of Environmental Policy and Planning* 15(4), 493–511.

Zhu, X. and Oude Lansink, A. (2012). Impact of CAP subsidies on technical efficiency of crop farms in Germany, the Netherlands and Sweden. *Journal of Agricultural Economics* 61, 545–564.

Chapter 4

Agroforestry practices: riparian forest buffers and filter strips

Richard Schultz, Thomas Isenhart, William Beck, Tyler Groh and Morgan Davis, Iowa State University, USA

1 Introduction

Intensive agriculture as practiced in much of the Temperate Zone around the world is not very friendly to the environment. Non-point source (NPS) pollution from this kind of agriculture has created major water quality issues for surface waters that originate or flow through these areas (Veum et al., 2009). In many landscapes in the Midwestern United States, more than 85% of the land is devoted to row crop agriculture or intensive grazing (Burkart et al., 1994). Small farms continue to be consolidated into larger farms in response to the need for economies of scale. In states east of the Rocky Mountains, vast areas are used to produce wheat, maize, soybeans and sorghum and to graze cattle (NRCS-USDA). Farm equipment that is operated by one person continues to become more sophisticated and able to cultivate and harvest larger and larger fields. The cost of the equipment and of labour further supports the continued expansion of large crop fields which are dependent on significant use of fertilizers and pesticides to optimize yields. In addition, to diversify income streams, farmers may fence off areas such as those along tightly meandering streams that are not suited to intensive crop production and graze livestock that usually have access to the streams within the fenced pastures. Livestock access to streams can do major damage to streambanks and stream water quality.

http://dx.doi.org/10.19103/AS.2018.0041.01

These trends of more intensive use of all available land in agricultural regions are likely to continue with the growing world population. Increased surface run-off laden with sediment and agrochemicals and streambank collapse continue to provide higher and more frequent peak stream flows. These are characterized by high sediment and agrochemical loads that result in more flooding, incision and widening of stream channels, reduction of base flows and reduction in water quality and the quality of aquatic ecosystems.

Despite our best efforts, it is unlikely that significant reduction in nutrient and sediment loading to surface waters will be achieved through voluntary, traditional in-field management alone (Dinnes et al., 2002). Increased use, by some farmers, of techniques such as cover crops, frequent side-dressing of small amounts of fertilizer, slow release fertilizer to promote soil and water quality, nutrient cycling efficiency and crop productivity have been studied as a way to reduce nutrient and sediment loading to surface waters and some farmers are using them (Snapp et al., 2005). However, they have some disadvantages, including increased farming costs, delay of spring soil warming and making it more difficult to predict nitrogen (N) mineralization, creating challenges to widespread adoption of such practices (Roesch-McNally et al., 2017).

The Natural Resources Conservation Service (NRCS) of the United States Department of Agriculture (USDA) National Water Quality Initiative is designed to provide both in-field and edge-of-field practices such as buffer and filter strips to promote soil health, reduce erosion and lessen nutrient run-off (https://www.nrcs.usda.gov/wps/portal/nrcs/detail/national/water/?cid=stelprdb1047761). One of the major NRCS Conservation Practice Standards is the Riparian Forest Buffer Standard (Practice 391). In those watersheds where the programme has been available, over 3600 farmers have taken advantage of the programme (Brewer, 2002). However, in the first 8 years of the programme, only 825 000 acres of farmland in the priority watersheds were enrolled with 11 impaired water bodies improved to the point of being removed from the US Environmental Protection Agency Impaired Waters List (USEPA 303(d) list). There are more than 390 million acres of cropland in the United States, many of them in need of conservation practices that protect surface waters (USDA 2012 Census of Agriculture).

2 Riparian forest buffers

Riparian forest buffers are an agroforestry practice that, when properly applied to the agricultural landscape, can enhance and diversify farm income opportunities, improve the environment and create wildlife habitat (Schultz et al., 2009). By developing an understanding of the interactions between the trees, shrubs and native prairie plants or introduced grasses, buffers can be designed to capture most surface run-off before it reaches the stream channel while also stabilizing the banks of the stream channel.

Riparian forest buffers are planned combinations of trees, shrubs, grasses, forbs and bioengineered structures adjacent to or within a stream channel designed to mitigate the impact of land use on the stream or creek. At the landscape level, riparian forest buffers link the land and aquatic environment and perform vital ecological functions as part of the network of watersheds that connect forests, prairies, agricultural and urban lands. By establishing and managing the trees, shrubs, grasses and forbs in the riparian zone, water quality and the aquatic ecosystems can be maintained or enhanced and the impact of floods can be mitigated. However, to be effective, riparian buffers must include plants that are adapted to the soils, topography and flood regime of the riparian zone and the stream as well as the long-term management by the landowner.

A well-established and maintained riparian forest buffer can:

- protect and improve water quality;
- stabilize eroding streambanks;
- help reduce flood impacts;
- recharge shallow groundwater;
- supply diverse food and cover for upland wildlife;
- enhance biodiversity of the landscape;
- improve carbon sequestration;
- improve aquatic habitats for fish and other organisms; and
- generate farm income from products harvested from the buffer.

An overview of the environmental benefits of riparian and other types of buffers has been provided by Lovell and Sullivan (2006) and Gundersen et al. (2010). The role of riparian and other agroforestry techniques in preventing nutrient run-off and NPS pollution is reviewed in Udawatta et al. (2002, 2006, 2011), Lee et al. (2003) and Simpkins et al. (2003). The impact of riparian buffers in enhancing water quality, preventing sediment trapping and streambank erosion is discussed in Zaimes et al. (2004), Liu et al. (2008), Klapproth and Johnson (2009a) and Udawatta et al. (2010). The broader potential contribution of riparian buffers and forests to carbon sequestration is reviewed in Dybala et al. (2019), while the biodiversity and broader social benefits they provide are discussed in Klapproth and Johnson (2009b,c).

Before designing and establishing a riparian buffer, it is critical to understand upland plant communities and their present management, the objectives of the landowner interested in installing a riparian forest buffer and their willingness and ability to manage that buffer.

Landowner concerns associated with establishment of buffers can include concerns such as:

- how much can buffers reduce sediment and nutrient movement into a stream;
- can buffers be used to heal gullies;
- can buffers reduce streambank erosion and slow stream meandering;
- what kind of buffer vegetation produces the best wildlife habitat and fishery;
- will trees in a riparian forest buffer fall into the stream and back up water into crop fields and field drainage tiles;
- are buffers a source of weed seeds;
- are cool-season grass filters as effective as riparian forest buffers;
- will forest buffers attract beavers that build dams that back up water;
- will deer become a problem for crops;
- how much maintenance is required to keep a buffer functioning properly;
- will a buffer be damaged by floods;
- is fencing needed to keep livestock out of a buffer;
- how much land will be taken out of crop production or pasture; and
- can specific products be harvested from the buffer to offset income losses from the land and similar other questions.

Riparian buffers can take different forms in response to landowner objectives and concerns as well as the regional location of the streams being buffered. Riparian forest buffers in agricultural landscapes in the Eastern United States, for example, may contain narrow corridors of remnant forests along streams with little else being needed to create an effective buffer of crop field run-off other than a grass filter lying between the crop field and the existing forest buffer. In the arid and semi-arid west, riparian forest buffers may consist of narrow strips of native flood plain species that often lie between grazed shrub and short grass communities and the stream. In the agricultural belt of the Midwestern United States, riparian forest buffers often need to be established from scratch.

Because riparian forest communities naturally evolved in the most fertile and moist position of the landscape, they are often easy to reestablish. However, in many agricultural landscapes, land uses have changed the hydrology so dramatically that the hydrology of these communities cannot be restored to their original condition. Stream channels have been incised and widened by higher discharge resulting from greater surface run-off from crop fields and heavily grazed pastures. Channelization of meandering streams, field tiling of some landscapes and urbanization also have contributed to higher stormflows and lower base flows. In many cases, water tables have been lowered to the point that the restored buffers require a plant community that did not naturally occur in that location. However, with proper planning and design the function of a healthy riparian forest community can be reestablished.

3 Riparian forest buffer design and function

Riparian forest buffers typically should be composed of three distinct management zones (Fig. 1):

- Zone 1: Undisturbed forest
- Zone 2: Managed forest and shrubs
- Zone 3: Run-off control (grasses and forbs)

These zones contain different kinds of plantings with different functions.

Zone 1 includes a zone of trees whose major function is to stabilize the streambank, provide a large long-term nutrient sink, help improve soil quality through annual leaf litterfall, provide vertical structure for wildlife habitat and potentially provide some shade to the stream channel to help stabilize daily stream water temperature especially if the desired fishery includes cold water demanding species such as trout (Table 1).

Trees should not be placed so close to the edge of the bank that they completely shade the stream throughout the whole day once they are mature.

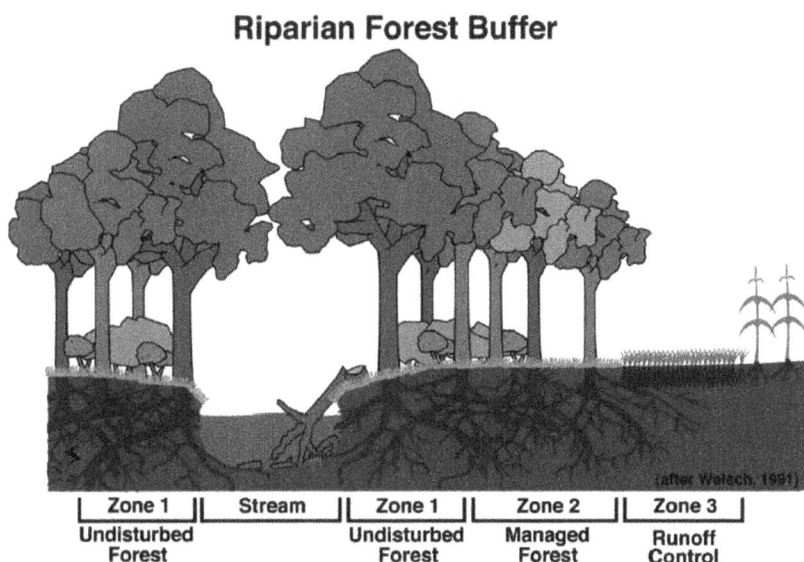

Riparian Forest Buffer

| Zone 1 | Stream | Zone 1 | Zone 2 | Zone 3 |
| Undisturbed Forest | | Undisturbed Forest | Managed Forest | Runoff Control |

Figure 1 The traditional three-zone riparian buffer. Source: reprinted with permission from Schultz et al. (2004) by American Society of Agronomy.

Table 1 Functions of the grass, shrub and tree components of riparian buffers

Kind of plant	Functions
Prairie grasses/forbs	1. Slow water entering the buffer
	2. Trap sediment and associated chemicals
	3. Add organic carbon to a range of soil depth
	4. Added carbon improves soil structure
	5. Improve infiltration capacity of the surface soil
	6. Above-ground nutrient sink needs annual harvest
	7. Provide diverse wildlife habitat
	8. Do not significantly shade the stream channel
	9. Provide only fine organic matter input to stream
	10. Can provide forage and other products
Shrubs	1. Multiple stems act as a trap for flood debris
	2. Provide woody roots for bank stabilization
	3. Litter fall helps improve surface soil quality
	4. Above-ground nutrient sink needs occasional harvest
	5. Adds vertical structure for wildlife habitat
	6. Do not significantly shade the stream channel
	7. Provide only fine organic matter input to stream
	8. Can provide ornamental products and berries
Trees	1. Strong, deep woody roots stabilize banks
	2. Litter fall helps improve surface soil quality
	3. Long-lived, large nutrient sink needs infrequent harvest
	4. Adds vertical structure for wildlife habitat
	5. Vertical structure may inhibit buffer use by grassland birds
	6. Shade stream, lowering temperature and stabilizing dissolved oxygen
	7. Provide both fine organic matter and large woody debris to the channel
	8. Can provide a wide variety of fibre products

The reason is that if the streambank is completely shaded, grasses needed to stabilize the bank will be difficult to establish and grow and bank erosion will continue with the chance that trees could fall into the channel creating problems. If, however, streams are large enough and there is a desire to create natural in-stream habitat, trees could be placed closer to the bank so that some large mature trees could fall into the channel to provide large woody debris that helps create important pool and woody in-stream habitat for a myriad of organisms. For this to happen, trees in Zone 1 would not be harvested during their life time. If in-stream large woody debris is not desired because of concern for development of log dams and increased water levels harvesting with replacement would need to be done.

Tree spacing in Zone 1 should also be wide enough to allow grass or some other kind of cover crop to grow under the trees. If that does not happen, bare soil will exist under the trees after the litter from the previous fall leaf drop has either decomposed or been washed into the stream by surface run-off or by floodwaters from out-of-bank stream flow. This is an important consideration since the riparian buffer has been designed to slow surface run-off by providing a frictional surface of plants completely covering the soil. If there is no permanent ground cover under the trees, soil erosion and even gullies will develop carrying sediment into the stream and weakening streambanks.

One other consideration for row spacing is the direction that the buffers run. Rows should be wider for tree and shrub rows that run east and west to allow a longer time for the sun to actually shine directly down to the soil level. Rows running north and south can be narrower because the sun shines down between the rows unimpeded by the shade from trees.

Species of trees to be planted in Zone 1 would depend on channel incision. If the channel is a natural channel that is in contact with its flood plain, meaning that flood events typically would occur once every 2–3 years, true riparian tree species should be selected because the water table in such a situation would be relatively close to the surface. If on the other hand the channel is deeply incised as happens in watersheds where flow regimes have changed because of land-use changes or climate change, both true riparian tree species and upland tree species could be planted. Upland species are often slower growing and longer-lived trees. Trees in this zone might also include fruit trees or even shrubs if there was a desire to provide a short-term income generating plant community. If the buffer was established with assistance from a government conservation programme, it may not be possible to sell harvested fruits until the enrolled programme has lapsed. Zone 1 is usually the widest zone occupying up to two-thirds of the width of the buffer.

Zone 2 combines planting of trees and shrubs. It both helps to manage floodwater and allows run-off to infiltrate or percolate into the soil so that waterborne nutrients or pollutants are absorbed and cleansed by the soil and vegetation. Zone 2 would be used if Zone 1 consisted of trees. Zone 2 would have several rows of shrubs again spaced far enough apart so that sunlight can reach the soil at least during part of the day.

These rows of multiple woody stems provide an important barrier for slowing floodwater that is moving out into the agricultural field and trapping debris brought by the floodwater. The debris may consist of a wide variety of objects including large woody debris that, if not trapped by the shrubs in Zone 2, would end up in the adjacent crop fields, physically damaging the crops and requiring time and money for the farmer to remove. The shrub species in this zone can consist of edible berries and/or decorative woody florals such as red osier dogwood and curly willow. These are valuable components of the floral and decorating industries and can thus provide the farmer with income. The shrubs in Zone 2 also can provide a significant wildlife benefit to the buffer especially in attracting birds that may be important in helping to control pests in the adjacent crop fields. Bird species that are attracted to the shrub zone can be manipulated through the selection of shrub species that are planted.

Zone 3 is the zone adjacent to the crop field and the most important of the three zones. The zone is designed to provide high infiltration, sediment trapping and nutrient uptake ability while also dispersing any concentrated flow that runs into it. Native grasses and forbs provide the best buffering. They help to restore biological and physical soil quality to heavily used soils by adding large amounts of carbon to the profile from rapid turnover of roots that contain more than 70% of the total biomass of native prairie plant communities. This carbon plays a key role in redeveloping soil macro-aggregate structure that helps facilitate the high infiltration rates needed to get surface run-off into the soil profile. The carbon also serves as a substrate for increased soil microbial activity that is important both in building soil structure and processing some agricultural chemicals that move in the surface and groundwater (Dornbush et al., 2008).

Cool-season grasses are good at protecting the soil because, when water runs through the filter, they lie down, allowing the water to run over them and protecting the soil. However, they do not slow the flow of the water and are thus better adapted for use in grass waterways. Native grasses and forbs slow the water because their stiff stems seldom bend in response to surface run-off. This results in the majority of the sediment in surface run-off being dropped on the crop field at the edge of the buffer prior to the water moving through Zone 3. That sediment can be moved back upslope where it can be used by crops. The mix of native prairie grasses and forbs provides excellent habitat for prairie and forest edge wildlife such as pheasants and quail. In regions where hunting game is an activity, riparian buffers with a Zone 3 prairie strip and a Zone 2 shrub strip can provide excellent bird hunting opportunities that some landowners lease to hunters.

If a crop field has more than a 5-8% slope, a pure switch grass (*Panicum virgatum*) strip can be planted at the field edge of Zone 1 to slow the water. The deep rooting habit of the native prairie grasses and forbs creates a soil that has high infiltration rates. Even on soils that have been cultivated for many years and lost their surface soil structure, native prairie plants can recreate soil structure and porosity similar to that of the original undisturbed soil in 8-10 years (Marquez et al., 2004, 2019). Cool-season grasses such as fescue and brome take a significantly longer time to improve infiltration rates to the same depth as under a native plant buffer community.

Planning considerations during buffer design should include a strong focus on the landowners' desires and objectives while also retaining the buffer's ability to provide critical environmental benefits and services. Buffers designed for stabilizing collapsing streambanks in deeply incised channels should have the first rows of trees set back far enough from the edge of the bank to allow grass or other dense ground cover to grow both above and on the bank in full sunlight. In such cases, trees that have a propensity for producing large major roots with many smaller fibrous roots should be selected as these species' root systems can provide the reinforcing structures that hold the banks in place. Buffers designed to maximize capture and filtration of crop field surface run-off should consist of native prairie grasses and forbs that provide stiff stems to slow water at the field edge of the buffer, dropping much of the sediment and then providing high infiltration rates to significant depths that allow the potentially nutrient laden water to be filtered through an active plant community root system.

4 Special design considerations and management

Design guidelines and planning tools for riparian forest buffers are provided by Bentrup (2008) and MacFarland (2017). While it is relatively straightforward to design a three-zone riparian buffer based on the above standard design, it is critical to fit that buffer to the actual landscape which often requires additional conservation practices that must be integrated with the buffer to make the system function to its maximum potential. To accomplish this integration of various potential conservation practices, design planning should include an on-the-ground walk-through with the landowner as well as an aerial photograph of the site that shows other conservation practices such as grass waterways and other problem areas such as field drainage tiles, gullies and areas of severe bank erosion and collapse. Recent advances in remote sensing have the potential to help buffer zone planning and management significantly (Herring et al., 2006; Goetz, 2006). Techniques such as high-resolution imaging and laser-based techniques can provide detailed information on buffer zone properties such as topography, buffer length, width and vegetation structure as well as stream flow.

As mentioned earlier, where there is severe bank erosion, Zone 1 trees should be set back from the bank edge to allow enough sunlight to support the growth of dense grass or other cover on the bank. Shrubs could replace trees in the first row or two of Zone 1 to reduce the potential of shading the stream. This is an important consideration in some prairie landscapes where warm-water streams exist and lowering water temperature is not desired. Replacing the first row or two of trees with shrubs may also be appropriate where there is significant landowner concern about large woody debris falling into the stream which might raise water levels, thus backing up water into field drainage tiles.

If the region includes field drainage tiles such as those found on the Des Moines lobe and other recently glaciated regions in the Midwestern United States, a grass waterway of introduced cool-season grasses should be planted over the tile as it passes under the buffer unless the tile can be replaced by a section of solid tile that has no access holes or cracks that provide potential access to plant roots. The deep roots of the native plant community or of trees will access the field drainage tile and plug it to the point that it will no longer carry water from the wet areas of the upland crop field.

In areas where grass waterways from the upland intersect the riparian forest buffer, Zone 3 of the buffer should be expanded out into the crop field or at the expense of the other two buffer zones. Grass waterways are designed to carry surface run-off water safely downhill. When this fast-paced water approaches a buffer strip, it must be dramatically slowed to allow the water to infiltrate into the soil below the buffer. In such cases, Zone 3 should consist only of native grasses and forbs with a strip of native grasses without forbs right at the field edge of Zone 3. The grass waterway should be widened at the edge of the buffer, creating a pyramidal structure with the base against the buffer to allow water a place to slow and sit before it moves through Zone 3.

Buffer widths can vary depending on space available, soil and slope conditions and landowner objectives (Fig. 2). Dosskey et al. (2015) have developed AgBufferBuilder, a GIS tool to design buffer strips using digital elevation models and buffer area ratio relationships, to develop buffers that have a constant level of trapping efficiency along the extent of the buffer. Riparian buffers as narrow as 10-15 m can provide surface erosion control, but nitrogen (N) reduction in subsurface flow may require widths of 30-46 m depending on the soil type and the slope of the riparian zone. When working with the landowner, it is important to determine the number of up and down field passes a field operative can make with the equipment available. In a rectangular field, buffer widths should allow the farmer to end tillage and harvesting passes up and down field in a way that will bring them back to the end of the field that they use for access to the field. If the stream meanders along the edge of the field, the buffer will need to vary in width to create the rectangular shape of the field.

The length of a riparian buffer system should ideally include both sides of the channel beginning in the headwaters of the watershed and extending continuously as far downstream as possible. Natural buffers can be part of the total length of a buffer system as long as they fall within the required widths needed to capture surface run-off and reduce the nutrient content of the subsurface flow to the channel. Leaving significant lengths of the channel without riparian buffers or some other kind of perennial cover can actually

Figure 2 A riparian buffer applied to the landscape. Note the variation in width of Zone 3, the native grass and forb zone, to fit the landowners' objectives. Also note the gentle smooth edge of Zone 3 along the buffer to allow ease of cultivation with large equipment.

create more problems for channel stability. If bank collapse takes place along unbuffered reaches of the channel, channel widths increase in that zone which causes water depths in the channel to decrease until the water hits a narrow channel that is stable because it is buffered. The turbulence caused by forcing the water into a narrower channel can increase undercutting and scouring along the buffered bank, causing it to collapse especially if the riparian buffer is relatively young.

Long-term management of the buffer is required to maintain its design functionality. No grazing should be allowed in Zones 1 and 2. If properly managed, flash or rotational grazing in dry soil conditions can be undertaken in a Zone 3 sown with cool-season grass. Some harvesting could be done in Zone 1 if the species used are stump or root sprouting species. If they are root sprouting species, row definition will be lost and woody tree stems could sprout into the Zone 3 grasses and forbs because tree roots extend laterally away from the bole to an average distance of one tree height from the base. Berries and shrub stems, to be used for ornamental purposes, can be harvested from Zone 2. Zone 3 cool-season grass could be cut for hay once in the growing season – ideally done in the season with the least potential flooding. More importantly, if Zone 3 consists of native grasses and forbs, it must be burned every 3–5 years to maintain the biodiversity of the plant community. If that is not done, invasive weeds or grasses, such as reed canary grass (*Phalaris arundinacea*), will find their way into the zone over time.

Riparian forest buffers may need to be used in conjunction with other riparian management practices such as streambank bioengineering, in-stream boulder weirs or constructed wetlands. Both streambank bioengineering and in-stream boulder weirs (Fig. 3) are designed to stabilize the channel by creating steps in the channel evolution process. In-stream boulder weirs are designed to slow down-cutting in the channel by creating a series of weirs in the channel with 1:4 slopes upstream of a series of cross-channel crest stones, and 1:20 slopes of rock downstream of the crest stones. Weirs are placed so that the upstream pool behind the crest stones of one weir are backed-up to the gentle downstream side of the weir upstream of that weir during base-flow. The goal is to reduce down-cutting and pool development downstream of the weir crest stones.

Streambank bioengineering is designed to stabilize eroding streambanks usually associated with a channel that needs to adjust to changes in discharge.

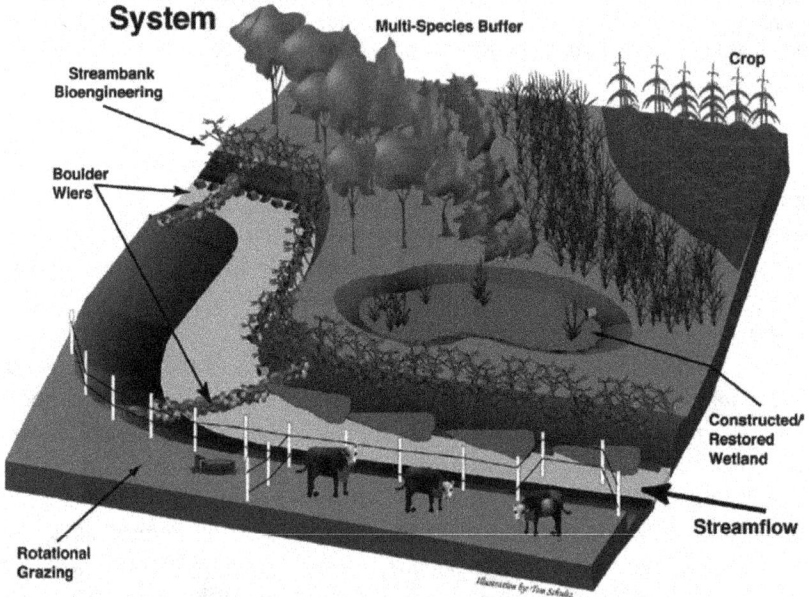

Figure 3 Riparian management system practices including from top right: streambank bioengineering, in-stream boulder weir structures, intensive rotational grazing, constructed wetlands and riparian forest buffer that could also be designed as a saturated buffer. Source: reprinted with permission from Schultz et al. (2004) by American Society of Agronomy.

Channels that are down-cutting, with the streambanks being taller than can be supported by the bank soil or parent material, can result in the banks collapsing, thus widening the channel. Bioengineering techniques include use of both hard engineering materials such as boulders for toe control and grasses and woody cuttings placed into the bank wall (Figs. 4 and 5).

Figure 4 Streambank bioengineering on streambanks that were severely eroded. Bear Creek in Central Iowa, USA – A USDA National Research and Demonstration Site – 1998.

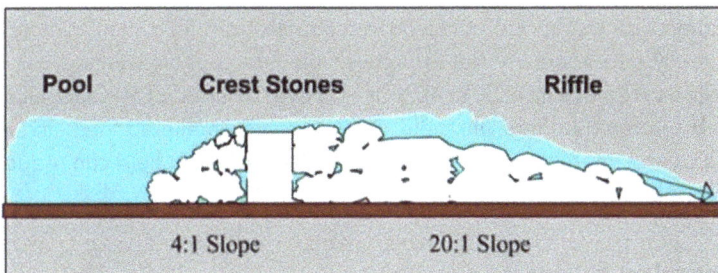

Figure 5 Boulder weir to control channel down-cutting by having a long gentle slope on the downstream side of the structure. Crest stones should be large stones that in the top layer allow fish to move over the crest. Boulder weirs are installed in a sequence with the upstream pool of the downstream weir backing water up to the long, gentle downstream rock sequence at base-flow conditions.

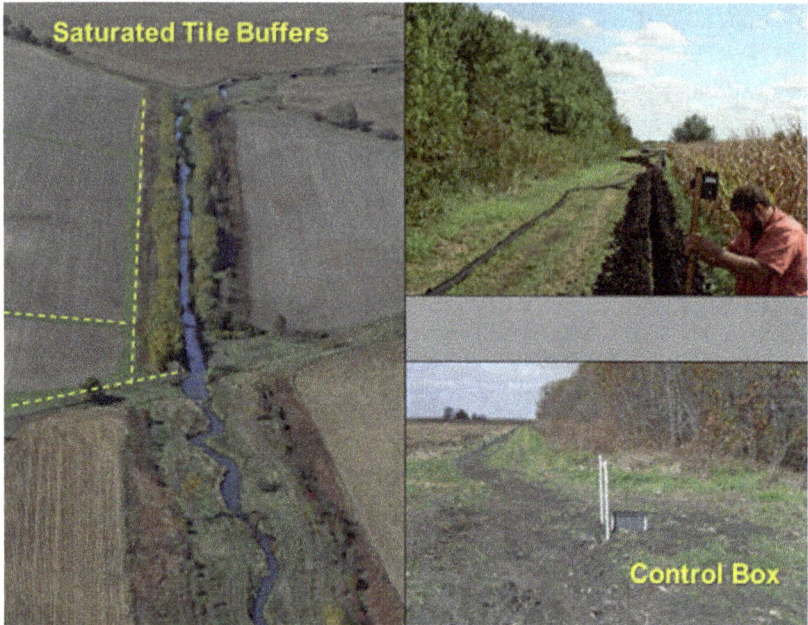

Figure 6 Saturated buffer. Field drain tile that intercepts tiles from upslope is laid parallel to the buffer. Water from the parallel tile now moves through the buffer instead of directly into the channel without any plant/soil treatment of the nutrient load.

Saturated riparian buffers are a relatively new addition to the buffer portfolio (Fig. 6). In this practice, a field tile that intercepts field tile draining upslope areas is laid parallel to the field edge of the buffer (Jaynes and Isenhart, 2014). Water flowing into this tile moves out of the tile and through the riparian buffer subjecting it to the treatment of the soil and plant community. Nitrate reduction is as high as 90% in the water flowing through the buffer before reaching the stream.

Streamside buffers cannot remove materials from field drainage tiles that exit directly into a stream. But an acre of tile-intercepting wetland has been calculated to remove from 20 to 40 tons of N over a period of 60 years. Likewise, creating saturated buffers with field tile that is laid parallel to the field side of the buffer and intercepts field tile that drains the adjacent field can reduce the nitrate content by 90% (Jaynes and Isenhart, 2014).

5 Assessing buffer performance

Tracer tests and isotope evidence shows that denitrification is the major groundwater nitrate removal mechanism in the buffer system (Schultz et al., 2004). Stratigraphy below buffers can determine the effectiveness of nutrient removal from shallow groundwater. With a shallow confining layer of till below a loamy root zone, buffers can remove up to 90% of the nitrate in groundwater. When the confining layer is found well below the rooting zone and porous sand and gravel are found between the till and the loam, residence time and contact with roots is dramatically reduced and buffers are unable to remove much nitrate from the groundwater. The difficulty in describing the stratigraphy below buffers makes it difficult to quantify the specific amount of remediation that a planned buffer might provide. To be able to measure in-stream water quality improvement, continuous buffers on both sides of the stream must extend at least 15 km.

Studies that have been conducted on the riparian forest buffers in the Bear Creek Watershed located on the Des Moines Lobe in Central Iowa in the United States have shown that a 7-m wide native grass filter strip on either side of the stream can reduce sediment loss to the stream by 95% and total nitrogen, phosphorous and nitrate and phosphate in the surface water by 60% (Schultz et al., 2004). This research suggests that adding a 9-m wide woody buffer to the grass filter results in removal of 97% of the sediment and 80% of the nutrients. There is also a 20% increase in removal of soluble nutrients with the added width (Simpkins et al., 2003; Lee et al., 2000, 2003). Riparian forest buffers can reach maximum efficiency for sediment removal in as little as 5 years and nutrient removal in as little as 10–15 years. Water can infiltrate in the soil up to five times faster in restored buffers in as little as 6 years after establishment compared to adjacent crop fields. Riparian buffer strips also have been shown to retain between 79% and 94% of the atrazine in run off from adjacent crop fields (Reungsang et al., 2005).

In terms of soil stability, it has been shown that buffered streambanks lose up to 80% less soil than row cropped or heavily grazed streambanks (Zaimes et al., 2004; Marquez et al., 2004). A study in the Central Claypan area of northeastern Missouri found that at the watershed scale, streambank erosion accounted for an average of 88% of the in-stream sediment and 23% of the N load on an annual basis suggesting the importance of using perennial vegetation to stabilize streambanks (Willett et al., 2012). Soils in riparian forest buffers contain up to 66% more total organic carbon in the top 50 cm than adjacent crop field soils (Tufekcioglu et al., 2003). *Populus* hybrids and switchgrass living and dead biomass sequester 3000 and 800 kg C ha^{-1} and immobilize 37 and 16 kg N ha^{-1}, respectively. Riparian forest buffers have more than eight times more below ground biomass than adjacent crop fields. Buffer soils show a 2.5-fold increase in soil microbial biomass and a fourfold increase in denitrification in the surface 50 cm of soil when compared with the adjacent crop field soil.

Bird species' use of buffers has shown that riparian forest buffers with a three-zone system of trees, shrubs and native grasses which provide a variety of habitat structure will support over 40 different bird species over the year in central Iowa compared to 8-10 species in non-buffer agricultural riparian zones with row crop culture to within 5 m of the channel (Berges et al., 2010). If properly designed, riparian buffers can protect a stream from chemical and sediment pollution while providing both terrestrial and aquatic wildlife habitat in agricultural landscapes that are dedicated to producing annual crops to feed humans and livestock or to create biofuels that replace fossil fuels.

6 References

Bentrup, G. 2008. Conservation buffers: design guidelines for buffers, corridors and greenways. USDA General Technical Report SRS-109. Available at: https://www.srs. fs.usda.gov/pubs/33522.

Berges, S. A., Schulte Moore, L. A., Isenhart, T. M. and Schultz, R. C. 2010. Bird species diversity in riparian buffers, row crop fields, and grazed pastures within agriculturally dominated watersheds. *Agroforest. Syst.* 79(1), 97-110. doi:10.1007/s10457-009-9270-6.

Brewer, M. 2002. Financial agents, water quality and riparian forest buffers. MS Thesis. Iowa State University, USA.

Burkart, M. R., Oberle, S. L., Hewitt, M. J. and Picklus, J. 1994. A framework for regional agroecosystems characterization using the national resources inventory. *J. Environ. Qual.* 23(5), 866-74. doi:10.2134/jeq1994.00472425002300050002x.

Dinnes, D., Karlen, D., Jaynes, D., Kaspar, T. C., Hatfield, J. L., Colvin, T. S. and Cambardella, C. A. 2002. Nitrogen management strategies to reduce leaching in tile-drained Midwestern soils. *Agron. J.* 94, 152-71.

Dornbush, M., Cambardella, C., Ingham, E. and Raich, J. 2008. A comparison of soil food webs beneath C3- and C4-dominated grasslands. *Biol. Fertil. Soils* 45(1), 73-81, doi:10.1007/s00374-008-0312-4.

Dosskey, M. G., Neelakantan, S., Mueller, T. G., Kellerman, T., Helmers, M. J. and Rienzi, E. 2015. AgBufferBuilder: A geographic information system (GIS) tool for precision design and performance assessment of filter strips. *J. Soil Water Conserv.* 70(4), 209-17.

Dybala, K. E., Matzek, V., Gardali, T. and Seavy, N. E. 2019. Carbon sequestration in riparian forests: a global synthesis and meta-analysis. *Glob. Change Biol.* 25(1), 57-67. doi:10.1111/gcb.14475.

Goetz, S. J. 2006. Remote sensing of riparian buffers: past progress and future prospects. *J. Amer. Water Res. Assoc.* 42(1), 133-43. doi:10.1111/j.1752-1688.2006.tb03829.x.

Gundersen, P., Lauren, A., Finer, L., Ring, E., Koivusalo, H., Saetersdal, M., Weslien, J. O., Sigurdsson, B. D., Högbom, L., Laine, J., et al. 2010. Environmental services provided from riparian forests in Nordic countries. *Ambio* 39(8), 555-66. doi:10.1007/s13280-010-0073-9.

Herring, J. P., Schultz, R. C. and Isenhart, T. M. 2006. Watershed-scale inventory of existing riparian buffers in northeast Missouri using GIS. *J. Amer. Water Res. Assoc.* 42(1), 145-55. doi:10.1111/j.1752-1688.2006.tb03830.x.

Jaynes, D. B. and Isenhart, T. M. 2014. Reconnecting tile drainage to riparian buffer hydrology for enhanced nitrate removal. *J. Environ. Qual.* 43(2), 631-8. doi:10.2134/jeq2013.08.0331.

Klapproth, J. and Johnson, J. 2009a. Understanding the science behind riparian forest buffers: effects on water quality. Virginia Cooperative Extension. Virginia Tech. Available at: https://pubs.ext.vt.edu/content/dam/pubs_ext_vt_edu/420/420-151/420-151_pdf.pdf.

Klapproth, J. and Johnson, J. 2009b. Understanding the science behind riparian forest buffers: effects on plant and animal communities. Virginia Cooperative Extension. Virginia Tech. Available at: http://pubs.ext.vt.edu/content/dam/pubs_ext_vt_edu/420/420-152/420-152_pdf.pdf.

Klapproth, J. and Johnson, J. 2009c. Understanding the science behind riparian forest buffers: benefits to communities and landowners. Virginia Cooperative Extension. Virginia Tech. Available at: https://pubs.ext.vt.edu/content/dam/pubs_ext_vt_edu/420/420-153/420-153_pdf.pdf.

Lee, K. H., Isenhart, T. M., Schultz, R. C. and Mickelson, S. K. 2000. Multispecies riparian buffers trap sediment and nutrients during rainfall simulations. *J. Environ. Qual.* 29(4), 1200–5. doi:10.2134/jeq2000.00472425002900040025x.

Lee, K. H., Isenhart, T. and Schultz, R. 2003. Sediment and nutrient removal in an established multi-species riparian buffer. *J. Soil Water Conserv.* 58, 1–8.

Liu, X., Zhang, X. and Zhang, M. 2008. Major factors affecting the efficacy of vegetated buffers on sediment trapping: a review and analysis. *J. Environ. Qual.* 37(5), 1667–74. doi:10.2134/jeq2007.0437.

Lovell, S. T. and Sullivan, W. C. 2006. Environmental benefits of conservation buffers in the United States: evidence, promise and open questions. *Agric. Ecosyst. Environ.* 112(4), 249–60. doi:10.1016/j.agee.2005.08.002.

MacFarland, K., Straight, R. and Dosskey, M. 2017. Riparian forest buffers: an agroforestry practice. Available at: https://www.fs.usda.gov/nac/documents/agroforestrynotes/an49rfb01.pdf.

Marquez, C. O., Garcia, V. J., Cambardella, C. A., Schultz, R. C. and Isenhart, T. M. 2004. Aggregate-size stability distribution and soil stability. *Soil Sci. Soc. Am. J.* 68(3), 725–35. doi:10.2136/sssaj2004.7250.

Marquez, C. O., Garcia, V. J., Schultz, R. C. and Isenhart, T. M. 2019. A conceptual framework to study soil aggregate dynamics. *Eur. J. Soil Sci.*, 1–14. doi:10.1111/ejss.12775.

NRCS Conservation Riparian Forest Buffer Practice Code 391. Available at: https://efotg.sc.egov.usda.gov/references/public/IA/Riparian_Forest_Buffer_391_STD_2014_05.pdf.

NRCS–USDA. The state of the land. Available at: https://www.nrcs.usda.gov/Internet/FSE_DOCUMENTS/nrcs143_012458.pdf.

Reungsang, A., Moorman, T. B. and Kanwar, R. S. 2005. Prediction of atrazine fate in riparian buffer strips soils using the root zone water quality model. *J. Water. Environ. Tech.* 3(2), 209–22. doi:10.2965/jwet.2005.209.

Roesch-McNally, G., Basche, A., Arbuckle, J., Tyndall, J. C., Miguez, F. E., Bowman, T. and Clay, R. 2017. The trouble with cover cropping: farmers' experiences with overcoming barriers to adoption. *Renew. Agric. Food Syst.* 33(4), 322–33.

Schultz, R. C., Isenhart, T. M., Simpkins, W. W. and Colletti, J. P. 2004. Riparian forest buffers in agroecosystems – lessons learned from the Bear Creek Watershed, Central Iowa, U.S.A. *Agrofor. Syst.* 63, 35–50.

Schultz, R. C., Isenhart, T. M., Colletti, J. P., Simpkins, W. W., Udawatta, R. P. and Schultz, P. L. 2009. Riparian and upland buffer practices. In: Garett, H. (Ed.), *North American*

Agroforestry: an Integrated Science and Practice. American Society of Agronomy, Madison, WI.

Simpkins, W., Wineland, T., Isenhart, T. and Schultz, R. 2003. Hydrologic setting control NO3-N removal in groundwater beneath multi-species riparian buffers. In: Kolpin, D. and Williams, J. (Eds), *Agricultural Hydrology and Water Quality: AWRA 2003 Spring Conference Proceedings*. American Water Resources Association, Middleburg, VA.

Snapp, S., Swinton, S., Labarta, R., Mutch, D., Black, J. R., Leep, R., Nyiraneza, J. and O'Neil, K. 2005. Evaluating cover crops for benefits, costs and performance within cropping system niches. *Agron. J.* 97(1), 322–32.

Tufekcioglu, A., Raich, J. W., Isenhart, T. M. and Schultz, R. C. 2003. Biomass carbon and nitrogen dynamics of multi-species riparian buffer zones within an agricultural watershed in Iowa, USA. *Agrofor. Syst.* 57(3), 187–98. doi:10.1023/A:1024898615284.

Udawatta, R. P., Krstansky, J. J., Henderson, G. S. and Garrett, H. E. 2002. Agroforestry practices, runoff and nutrient loss: a paired watershed comparison. *J. Environ. Qual.* 31(4), 1214–25.

Udawatta, R. P., Motavalli, P. P., Garrett, H. E. and Krstansky, J. J. 2006. Nitrogen losses in runoff from three adjacent agricultural watersheds with claypan soils. *Agric. Ecosyst. Environ.* 117(1), 39–48. doi:10.1016/j.agee.2006.03.002.

Udawatta, R. P., Garrett, H. E. and Kallenbach, R. L. 2010. Agroforestry and grass buffer effects on water quality in grazed pastures. *Agrofor. Syst.* 79(1), 81–7. doi:10.1007/s10457-010-9288-9.

Udawatta, R. P., Garrett, H. E. and Kallenbach, R. 2011. Agroforestry buffers for non-point source pollution reductions from agricultural watersheds. *J. Environ. Qual.* 40(3), 800–6. doi:10.2134/jeq2010.0168.

USEPA 303(d) List. https://www.epa.gov/tmdl/program-overview-303d-listing-impaired-waters.

Veum, K. S., Goyne, K. W., Motavalli, P. P. and Udawatta, R. P. 2009. Runoff and dissolved organic carbon loss from a paired watershed study of three adjacent agricultural watersheds. *Agric. Ecosyst. Environ.* 130(3–4), 115–22. doi:10.1016/j.agee.2008.12.006.

Willet, C. D., Lerch, R. N., Schultz, R. C., Berges, S. A., Peacher, R. D. and Isenhart, T. M. 2012. Streambank ersion in two watersheds of the Central Claypan Region of Missouri, United States. *J. Soil Water Conserv.* 67(4), 249–63. doi:10.2489/jswc.67.4.249.

Zaimes, G., Schultz, R. and Isenhart, T. 2004. Streambank erosion adjacent to riparian forest buffers, row-cropped fields and continuously-grazed pastures along Bear Creek in Central Iowa. *J. Soil Water Conserv.* 59, 19–27.

www.ingramcontent.com/pod-product-compliance
Lightning Source LLC
Chambersburg PA
CBHW050538270326
41926CB00015B/3288